STUDENT UNIT GUIDE

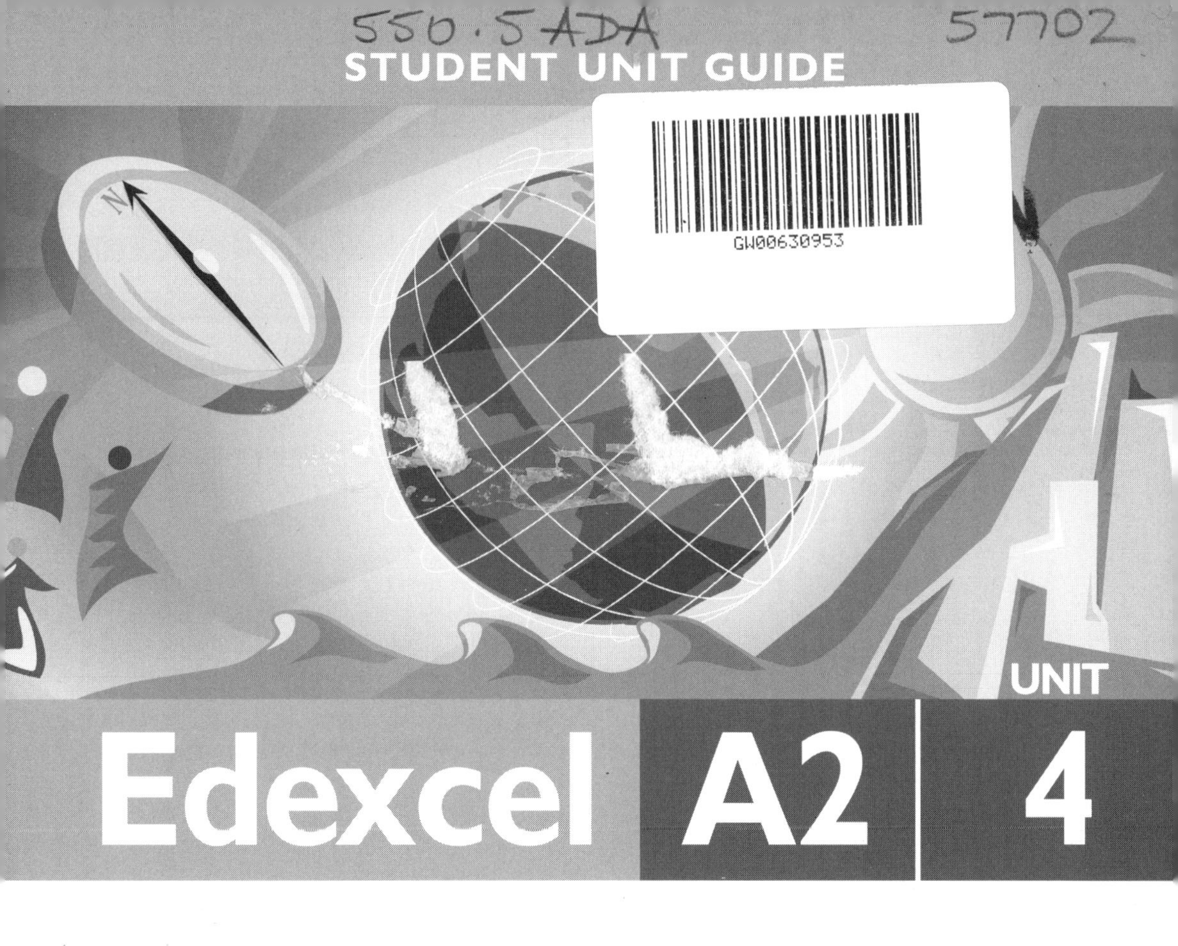

Edexcel A2 | 4

Geography

Geographical Research

Kim Adams and David Holmes

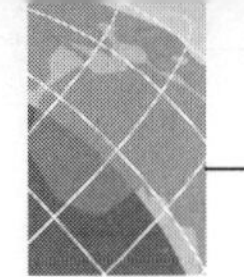

Philip Allan Updates, an imprint of Hodder Education, an Hachette UK company, Market Place, Deddington, Oxfordshire OX15 0SE

Orders

Bookpoint Ltd, 130 Milton Park, Abingdon, Oxfordshire OX14 4SB
tel: 01235 827720
fax: 01235 400454
e-mail: uk.orders@bookpoint.co.uk
Lines are open 9.00 a.m.–5.00 p.m., Monday to Saturday, with a 24-hour message answering service. You can also order through the Philip Allan Updates website: www.philipallan.co.uk

ISBN 978-0-340-99084-1

First printed 2009
Impression number 5 4 3 2 1
Year 2014 2013 2012 2011 2010 2009

This Guide has been written specifically to support students preparing for the Edexcel A2 Geography Unit 4 examination. The content has been neither approved nor endorsed by Edexcel and remains the sole responsibility of the authors.

Typeset by Philip Allan Updates

Printed by MPG Books, Bodmin

Contents

Introduction

About this guide

This guide is for students following the Edexcel A2 Geography course. It aims to guide you through Unit 4: Geographical Research.

This **Introduction** provides an overview of the unit, introduces the specification content and gives tips on carrying out research and tackling the exam.

The **Content Guidance** section provides a detailed guide to the six options from which you will choose one to specialise in. Research activities and tips are also given.

The **Question and Answer** section provides samples of different grades of work in the final essay report with examiner's comments on how to improve your performance in the exam.

Overview of Unit 4

For Unit 4: Geographical Research you must choose **one** from six available options. In the exam you have to write one report-style essay in 1.5 hours. Some options are more physically biased, some more human.

The options (starting with the more physically based options and moving towards the more human based options) are as follows:

Option 1 Tectonic activity and hazards

Option 2 Cold environments: landscapes and change

Option 3 Life on the margins: the food supply problem

Option 5 Pollution and human health at risk

Option 6 Consuming the rural landscape: leisure and tourism

Option 4 The world of cultural diversity

In the specification, each option is divided into four enquiry questions, and these are then divided into four sub-enquiry topics. An example based on Option 2 is shown in Table A.

Questions in the final exam may be based mainly on one of the four enquiry questions, but are more likely to be based across several. Your work on this unit may be thought of as a progression along a **timeline** (see Table B). The important thing throughout this sustained piece of work is to create an orderly set of notes and annotated articles so that when you find out the overall focus of the real exam a few weeks in advance, you have identifiable sections to concentrate on, a trail of extra references/sources to check concepts and improve factual details, and, if relevant, additional topical examples.

Table A Cold environments: landscapes and change

Main enquiry questions	1 Defining and locating cold environments	2 Climatic processes and their causes	3 Distinctive landforms and landscapes	4 Challenges and opportunities
Sub-enquiry topics	• Cold, glacial, periglacial environments • Systems • Distribution changes over time • British Isles glacial and interglacial history	• Climatic causes • Long-term changes • Meteorological processes • British Isles spatial and temporal relationships between glacial and periglacial environments	• Glacial geomorpho-logical processes • Glacial landforms • Periglacial geomorpho-logical processes • Periglacial landforms	• Definitions and links • Past and present challenges and opportunities • Overcoming challenges • Approaches to use and management

Table B Timeline

Decision on which option Approx 10 weeks' class work and homework, researching into each of the sub-enquiry questions for ONE option	• Your notes should make clear reference to the main four enquiry questions. Build up a series of detailed case studies and smaller examples so that you can be flexible in the final exam. There is as much skill in rejecting information that you know as there is in selecting the most appropriate bits. • Summarise information in factfiles, glossaries, spider topic webs. • Each example you research should have information on the key synoptic elements of this unit: people/places/power. • Before the final exam you should write at least five practice essay/reports, some under timed conditions, with/without your notes. Create a glossary of key terms and learn them.
The pre-release research focus, given at least 4 weeks before the actual exam	• Look closely at the research focus, it gives major clues as to the final title in the exam. • Practise different command words, such as : – Discuss… – To what extent… – Explain why or how… • Ensure you understand key terms like process, factor, players, geographical, challenges, responses, causes, impacts. • Make short summaries on cards of key facts to learn for the exam. • Learn your glossary – geographical vocabulary is essential.
The exam 90 minutes	• You will have a page in your exam booklet devoted to making a plan – use it. It doesn't matter how messy and scribbled this is, it's for you not the examiner. Students who spend 5 minutes or so on a plan usually increase their final marks substantially. • Then get writing, remembering the outline of how your essay will be marked (see Exam tips: DRACQ). • Try to finish the essay/report – even at the risk of making it obvious that you ran out of time by leaving a gap and going to the conclusion. Otherwise you throw away valuable marks.

Case studies and examples

Develop a range of mini-examples and more detailed case studies as you progress through the enquiry questions. Create mini-factfiles ensuring you cover the **synoptic** element of the specification: places/people/power. More detailed checklists for assembling notes under the enquiry questions for each option may be found in the Content Guidance section.

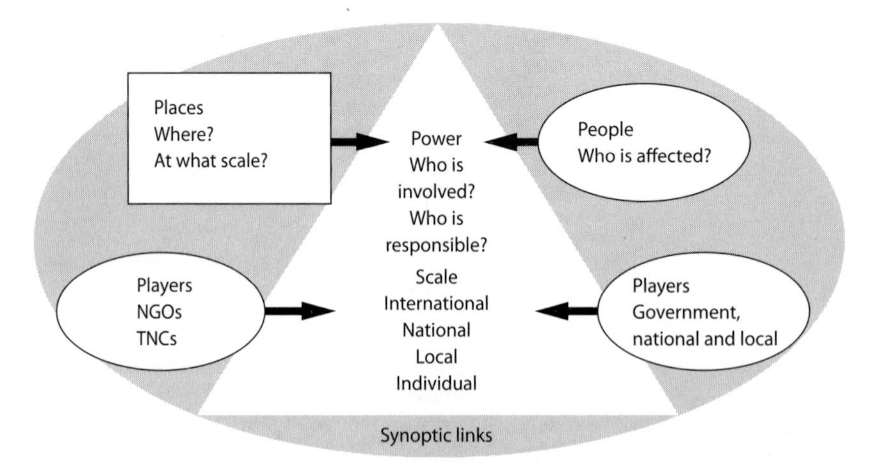

Places, people, players and power

Research tips

Researching an extended project, long essay or coursework can easily get out of control. Here are some strategies to stop that happening.

Planning: design a structure before you start

- Be clear about what you are being asked to do.
- Plan your work — divide it into clearly defined sections, using the enquiry questions with subdivisions *or* key case studies. Set up this structure so that it is repeated wherever you store information: your folder, your bookmarks etc.
- Use a topic map of the specification and the main case studies as a contents list.
- Work out a timescale for the work.

Reading/research

As you research material, constantly remind yourself of your aims, title etc. Ask yourself:

- Is this relevant?
- How does this answer my topic?
- Why should I include this?
- What information do I still need to find?

Share ideas and research with your friends. Writing a short glossary as you go will also be invaluable for final revision.

Sources of information and taking notes

- Beware bias in any resource. For example, when using a website: who owns it; is it authentic; is it up to date? Be careful and selective with Wikipedia and YouTube in particular.
- Keep a record of where your notes/articles originate from — you may need to return to them during revision.
- Skim-read all sources to get a feel for content, then reject/accept accordingly.
- Keep it relevant — ensure any notes relate specifically back to the specification to reduce your workload.
- If material overlaps different key questions on the specification, note in the margin/header/footer other areas supported.
- Do not plagiarise — put information into your own words and never just copy and paste information.

Possible sources of information include: magazine articles, pamphlets, textbooks, GIS systems, the internet (e.g. internet gateways, metasearch engines, local and national authority and international authority websites, online newspapers and books), fieldwork, geography department videos and DVDs.

Exam tips

How to use the mark scheme

Part of the mark scheme is shown below. It is **generic** because it applies to any exam question. The mark scheme sections are known as DRACQ:

D = Discuss/define

R = Research

A = Application

C = Conclusions

Q = Quality of written communication

Only the top level for each of the DRACQ criteria is reproduced here. You should get a full mark scheme from the Edexcel website **www.edexcel.com** to see how to maximise your learning and skills in the final exam.

Some students find it useful to remind themselves of the DRACQ on their planning page. You will see that writing a full introduction and a thorough conclusion with sub evaluations in the report will gain you 20 of the available 70 marks. The application criteria are the most critical — as always in exams it is not so much what you know but how you use it.

The only way to get high marks is to practise writing essay reports: first as research tasks, probably longer than you would be able to write in the exam, and then under timed conditions.

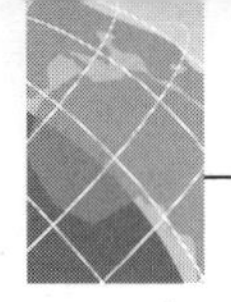

Table C Part of the generic mark scheme

	Introducing, defining and focusing on the question — 10 marks	
D	9–10	• Clear reference to title — develops a focus • Ongoing evaluation throughout report • Understands the complexity of the question
	Researching and methodology — 15 marks	
R	12–15	• Wide range of relevant case studies used (by scale and/or location) • Relevant concepts, and/or theories used • Factual, topical evidence • Indication of methodology, i.e. how evidence was sampled/selected
	Analysis, application and understanding — 20 marks	
A	17–20	• All research applied directly to question set • High conceptual understanding • Cogent argument • Appreciation of different values/perspectives about the question • Any maps/diagrams are used to support answer
	Conclusions and evaluation — 15 marks	
C	12–15	• Clearly stated • Thorough recall of content/case studies used in essay
	Quality of written communication and sourcing — 10 marks	
Q	9–10	• Coherent structure and sequencing with obvious report style sub-sections • Excellent standards of spelling and punctuation • Geographical vocabulary used correctly • Diagrams/maps, if used, incorporated into text and support argument • Referenced/acknowledged material: obvious evidencing/sourcing from wide range of sources (texts, journals, internet, DVDs etc.)

Getting the best out of the pre-release research focus

The following format will be used in the pre-release research focus:

- **Explore...** This will guide you to the enquiry question/s and sub-questions chosen by the principal examiner. It refers to the background concepts, processes, theories and models involved, where relevant.
- **Research...** This will hint at the type of supporting evidence needed, for example a range of areas/types/strategies, and means a focus on the geographical places, case studies and examples illustrating these.

Here is an example from Option 5 Pollution and human health at risk.

Pre-release focus

- **Explore** the varying links between the cause and spread of health risk and geographical features.
- **Research** a range of types of health risk and location.

Unpicking the research focus

You could make a table and take a systematic approach, as shown in Table D.

Table D

Pose a question from the research focus	Which enquiry questions?	Details	Actual case study you could use/need more research on
What are health risks? Range needed?	EQ1: Types, patterns, globally, over time	Range: chronic, infectious, traumas	Range, e.g. by scales: small–global Economic pathway: MEDC–NIC–LEDC
What are the causes? What are the geographical features? Spread?	EQ2: Complex causes, relationship to socioeconomic status EQ2: Links between some diseases and geographical features at variety of scales. Development and spread Model: diffusion EQ3: Pollution Also a bit of EQ4 on management, which may reduce worst effects/increase problem if mismanaged	**Human** • Infrastructure, transport • Population concentrations, urban focus? • Lifestyle choices, economic development • Poverty • Type of industry/employment and role of pollution, e.g. radiation, chemical spills	• Swine flu 2009 — air travel, intensive pig farming, some mismanagement by WHO and Mexico authorities at start of pandemic; use relocation diffusion model • Chernobyl, nuclear-radiation • Lifestyle feature: obesity and fat camps of UK, USA, China • Cholera: political corruption and famine vulnerability — Zimbabwe 2008
		Physical • Role of water especially rivers • Role of climate change and spread of some risks	• Cholera refugee camps Darfur 2006 • Malaria Kenya twenty-first century
Varying links?	Means some links are strong, some weaker	Strong — possibly direct Weaker	Shanty town conditions, industrial pollution Political and social reasons, e.g. HIV/AIDs?

Tackling exam questions: command words and key words

Every examination question in this unit will have the following components:

- **Command word/s** — e.g. evaluate, discuss
- **Key words** — for focus of the discussion, e.g. challenge, relationship, spatially
- **Geographical focus** — a focus for your discussion of the question, e.g. food security issues, responses to tectonic hazards

Table E gives a list of common command words used in exam questions.

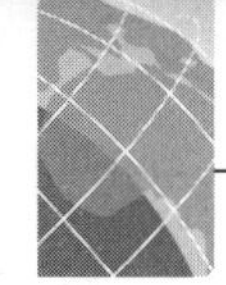

Table E Common command words

Command word	Meaning
Assess	Weigh up both sides of an issue/solution and come to a conclusion
Compare	Set items side-by-side and identify similarities and differences
Consider	Describe and explain a range of different views on a subject; explore a range of positions
Contrast	Point out only the differences between two or more items
Discuss	Give both sides of an argument (for and against), and come to a conclusion
Evaluate	Weigh up several options or arguments and come to a conclusion about their importance/success
Examine	Investigate in detail; offer evidence for and against
Explain	Provide a detailed set of reasons why something is like it is
Justify	Give the reasons why something should be done; and why other options should not
Suggest reasons/how	Provide an explanation; say why
To what extent	Say 'how far' you agree with a statement or option by examining its advantages and disadvantages

Your essay report needs to show that you can both develop an argument and sustain it. Plain descriptions of case studies will not win you many marks at A2, but using an evaluative style will get you into the highest levels. An evaluative answer explores several sides of an argument, and shows that geography is complex and there is no 'one' solution or viewpoint. To help develop an evaluative style try to use some of these signposting and linkage terms:

however, but, although, nevertheless, on the other hand, conversely, despite, whereas

Key words are instructions that tell you what to write your answer about. They provide you with the *focus* of the question. Some common key words are given in Table F.

Table F Common key words

Key word	Meaning
Action	Strategies and methods that might be used to manage a problem
Causes	The reasons why something happens
Challenges	Difficult, large-scale problems that require solutions
Changes	Transformations that take place over time
Characteristics	The key features of something, often split into: physical/environmental and social/economic in geography
Concerns	Aspects of an issue that are worrying

Key word	Meaning
Conflicts	The issue over which two or more groups are arguing
Consequences	The results of a change or process; they can be positive or negative
Decision maker(s)	Individuals, groups and organisations with some influence over how a problem should be tackled
Distribution	The geographical pattern, most often on a map
Effects	The results of a process occurring (comes after causes)
Factors	The underlying causes of a problem or process
Futures	The range of pathways humans could choose to take, e.g. sustainable or business-as-usual
Impacts	The results of a process or change on people and the environment; can be positive and negative
Implications	The likely consequences/impacts of a change
Interrelationships	Links between two or more features; changing one feature leads to changes in others
Issues	Concerns; problems that are worrying
Management	Using policies and strategies to minimise or reduce problems
Pattern(s)	The distribution of something; where things are; most often on a map
Places	Areas of the world, of varying scales
Player	An individual, group or organisation involved in an issue (stakeholder)
Power	Amount and importance of influence an individual/group or organisation may have
Problems	Issues that worry people; the negative results of a process or change
Processes	Sequences of events that cause change to take place
Relationships	Usually used to mean the links between a cause and its effects
Responses	Ways in which people might react to a problem or change
Role	The functions or activities of an individual, group or organisation
Scale(s)	The size of a feature — global, regional, national, local
Spatial	Variation in space (across an area)
Strategy	A method used to manage a problem or issue
Structure	How parts of something are arranged in relation to each other; the links between the parts
Temporal	Variation in time (change over time)
Threats	The causes of negative impacts
Variation	How far something differs from the norm or the average; include anomalies (differences from the norm)
Ways	Actions or strategies that might be used to deal with a problem or issue

Tackling exam questions: planning your answer

There are many ways you could approach planning your answer:

- a basic bullet list or a spider diagram
- a spectrum diagram: where you consider two sides of a question: large scale versus small scale, strengths versus weaknesses, things I agree/disagree with etc.
- DRACQ (see above): plan you answer according to the mark scheme
- BUG — **B**ox the command word(s), **U**nderline the key words, **G**lance back to ensure you understand the whole question and choose case studies/examples
- Note down the timings for your answer, e.g. 5 minutes to plan answer, 10 minutes to write introduction, 50–60 minutes for main section of report discussing title and sub-conclusions, 15 minutes for final conclusion and final check

Tackling exam questions: your introduction

- Identify the command words and respond to them — not to what you did in a mock or practice question at school/college.
- Introduce by focused discussion not by a list of definitions — do not use a pre-learnt introduction.
- Introduce your case studies and a model, if relevant, to answer the question effectively. Avoid too much detail at this stage.
- Start using the correct geographical terminology.
- Start arguing a case if that is what the question asks, but do not give away your final conclusion. You may have changed your mind after looking at your case studies.

Tackling exam questions: your analysis section

- Examiners like diagrams that support/extend knowledge. Make sure any diagram is relevant to the title.
- Maps just showing location are a waste of time: the examiner will know where something is.
- Link the theoretical concepts and models to reality/case studies.
- Try to link up your paragraphs — see the signposting and linkage terms above.
- Remember to link each case study back to the precise title. Try to argue a case — the titles are designed to allow you to do this.

Remember that all important **conclusion**. It summarises main points/arguments and should have been built up progressively in the report as sub-evaluations.

Summary

Below is a summary of how you can succeed in your Unit 4 Geography exam.

S — **Synthesis** — use all your skills as a geographer
U — **Understand** what is required
C — act on **Command** words
C — remember **Case studies** and **Concepts** — appropriate and relevant
E — **Exam question** — choose one you have prepared for
S — keep to a strict time **Schedule**
S — make full use of the mark **Scheme**

Content Guidance

Unit 4: Geographical Research consists of six research options. You are required to study one of the options, which is assessed by a 70-mark, 90-minute examination.

- Option 1 Tectonic activity and hazards
- Option 2 Cold environments: landscapes and change
- Option 3 Life on the margins: the food supply problem
- Option 4 The world of cultural diversity
- Option 5 Pollution and human health at risk
- Option 6 Consuming the rural landscape: leisure and tourism

The options in Unit 4 range from those with a strong physical focus (e.g. cold environments) to those concerned with more social and cultural geography. There is also common emphasis of people, places and power which runs through the global synoptic content of the unit. The research process is central to the successful completion of the unit, together with refined selection and synthesis skills.

In this Content Guidance section the information for each option is divided into sub-sections that match the enquiry questions given in the specification.

Tectonic activity and hazards

Introduction

Hazards, including tectonic activity, become disasters because of human vulnerability. An earthquake in an unpopulated area is not a disaster, because there is no human impact. Tectonic activity generates a wide range of natural hazards including lava flows, ash falls, ground shaking and tsunamis. The primary cause of these is plate tectonics — the movement of the crustal plates on the Earth's surface. As a result, tectonic hazards have a particular distribution concentrated on the plate margins and 'hot spots'.

Tectonic hazards can generate significant risk (short, medium and longer term) to people who live in tectonically active regions. Risks are closely related to:

- vulnerability of the population
- magnitude and frequency of hazard events
- level of economic development
- population density
- community preparedness/education of the people

People respond to hazard risks in a variety of ways, e.g. trying to prevent or modify the hazardous event, modifying the loss burden, and reducing human vulnerability. The level, scale and effectiveness of the response depend on available resources. Much current scientific research is focused on increasing our understanding of tectonic hazards and on improving our ability to cope with their impacts.

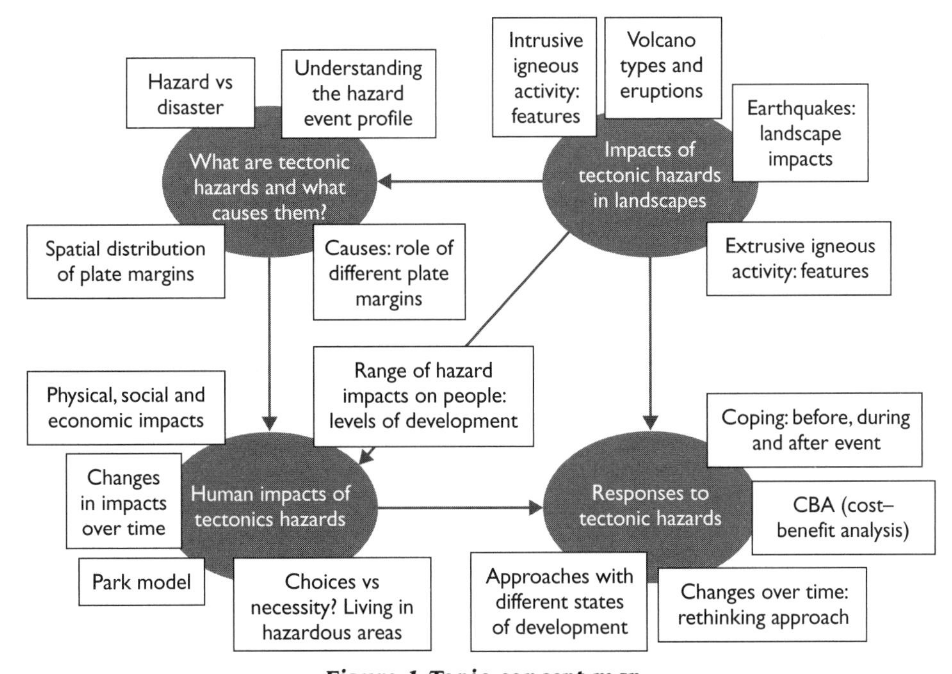

Figure 1 Topic concept map

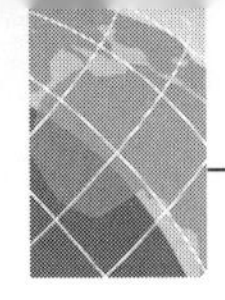

Research tip
While thinking of the synoptic aspects of the course (Figure 2), a simple checklist based on the topics within each enquiry question in the specification will help you to separate your research into manageable chunks. When you are researching tectonics, use a checklist linked to the specification (see Table 1). Remember to note the main reference source as this is a requirement for the exam — you are expected to quote some key sources: from books, newspapers, websites, DVDs, podcasts, pamphlets etc.

Figure 2 The synoptic context

Table 1 Checklist for tectonics

Enquiry question 1 **What are tectonic hazards and what causes them?**	**Enquiry question 2** **What impact does tectonic activity have on landscapes and why does this impact vary?**	**Enquiry question 3** **What impacts do tectonic hazards have on people and why do these impacts vary?**	**Enquiry question 4** **How do people cope with tectonic hazards and what are the issues for the future?**
• What is a tectonic event? • Hazard vs disaster • Global plate boundaries and types (transform, convergent etc.) • Patterns and distribution — comparisons of earthquakes vs volcanoes • Understanding the hazard profile (and key terms)	• How extrusive igneous activity varies spatially • Features linked to intrusive igneous activity • Formation and types of volcano • Plate boundaries vs hot spots • Earthquake impacts on landscapes and associated features	• Tectonic risk and people — influences • Perceived vs real tectonic risks • Comparison of risks in locations with different states of development • Impacts on people: social and economic • Hazards impacts are dynamic: long and shorter term changes	• Approaches to management: variations by place and level of economic development • Modifying event? • Modifying loss burden? • Modifying vulnerability? • Changes in approaches over time (before, during and after)
Key refs used	Key refs used	Key refs used	Key refs used

Research activity
Use Table 1 as an initial stimulus, and carry out your own research audit to obtain background sources of information. Make a note of any key references used.

Enquiry question 1

What are tectonic hazards and what causes them?

A **tectonic event** is a physical occurrence resulting from the movement or deformation of the Earth's crust. They are predominantly earthquakes or volcanic eruptions and their associated activities. Such events become **tectonic hazards** when they have the potential to cause loss of life and damage to property. But not all events are hazardous. Many earthquakes in active areas are low in magnitude and may occur deep below the surface. People living in the area may not even feel them. For example, in a typical week in April 2009 there were over 150 'earthquakes' in the area around Los Angeles, but virtually all were below 3.0 on the Richter scale. They presented no danger to the local people.

A hazard becomes a **tectonic disaster** when an event materialises and causes extensive destruction and a number of fatalities. For example, the eruption of Mt Pinatubo in the Philippines in 1991 killed about 300 people, over 80,000 homes were destroyed or damaged and 800 km² of agricultural land was buried beneath a thick layer of ash.

The **hazard event profile** is a diagram that tries to represent the characteristics (magnitude, speed of onset, duration, areal extent, etc.) of different types of hazards (Figure 3). These are useful for making comparisons between different types of tectonic event.

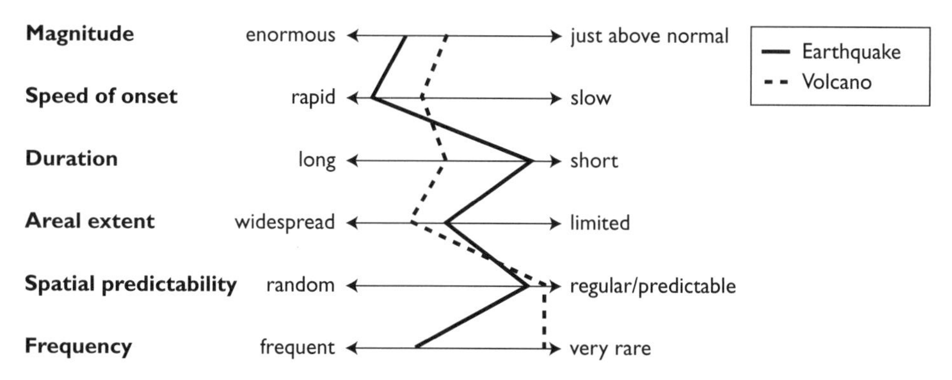

Figure 3 Typical hazard event profile for an earthquake (destructive boundary) and volcano

Causes of tectonic hazards

The tectonic plates of the Earth's crust move relative to each other so there is a slow build up of stress within rocks. When the pressure is released, parts of the Earth's surface experience an intense shaking motion, which typically only lasts a few seconds. This is the main cause of earthquakes. Earthquakes can be classified according to the depth of their focus, which affects the amount of surface damage that results from the event. Three broad categories are recognised:

- deep focus: 300–700 km deep
- intermediate focus: 70–300 km deep

- shallow focus 0–70 km deep. These cause the greatest degree of damage and account for approximately 75% of all earthquake types

A volcano is a landform that develops around a weakness in the Earth's crust. At this point, molten rock or magna is forced out or extruded. The nature of the volcanic eruption, i.e. viscosity (thickness and stickiness of the magma), the amount of dissolved gases and how easily they can escape, is determined by the tectonic context. Volcanic eruptions can be classified by the 'volcanic explosivity index' (VEI) where 0 = non-explosive and 8 is very large.

Table 2 Comparison of the distribution of volcanoes and earthquakes

Earthquakes	**Volcanoes**
Distribution tends to be tightly focused in linear bands (belts), but can occasionally occur away from these, e.g. New Madrid earthquakes (USA) in 1811–12 and Dudley (UK) earthquake 2002 (both examples of an intra-plate earthquake)	A good proportion of volcanic activity occurs at or near plate margins (inter-plate). Can be at a constructive margin: e.g. Iceland, or destructive: e.g. Andean volcanoes in Chile and Peru. Generally distribution is both *linear* and *clustered*
Most intense and frequent earthquakes are associated with subduction/destructive zones, and sometimes active conservative margins, e.g. San Andreas, USA	Volcanoes can also be found intra-plate at hot spots, e.g. Hawaii in the Pacific
High intensity shake events are *not* normally associated with non-active constructive (transform) boundaries	Continental margins are associated with the most violent volcanic eruptions (due to the presence of both oceanic and continental crust)

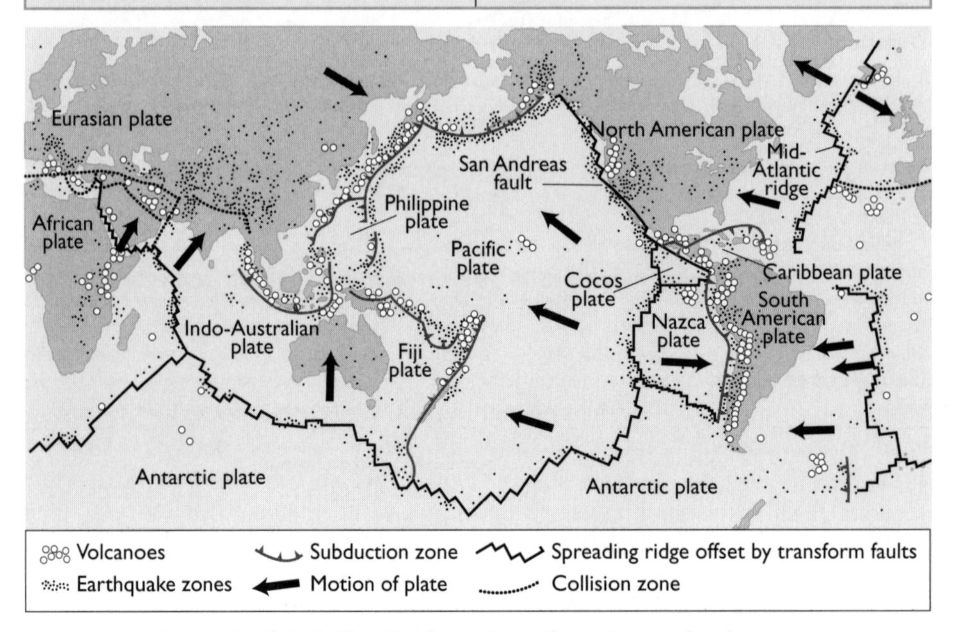

Figure 4 Global distribution of earthquakes and volcanoes

Tsunamis ('harbour waves') are waves caused by the rapid movement or deformation of the seabed. This motion can be initiated by undersea earthquakes, volcanoes, landslides or slumps. They are classified as secondary hazards since they are events caused by a tectonic event.

The **distribution** of earthquakes and volcanoes (Table 2, Figures 4 and 5) has a distinct geography. There are two important areas of activity: the oceanic fracture zone (OFZ) and the continental fracture zone (CFZ).

- The **OFZ** is a belt of activity through the oceans along the mid-oceanic ridges, coming ashore at Africa, the Red Sea, the Dead Sea Rift and California.
- The **CFZ** is a more important belt of activity (passing though more populous areas). It flows through the mountain ranges of Spain, via the Alps to the Middle East, to the Himalayas and the East Indies and then circumscribes the Pacific.

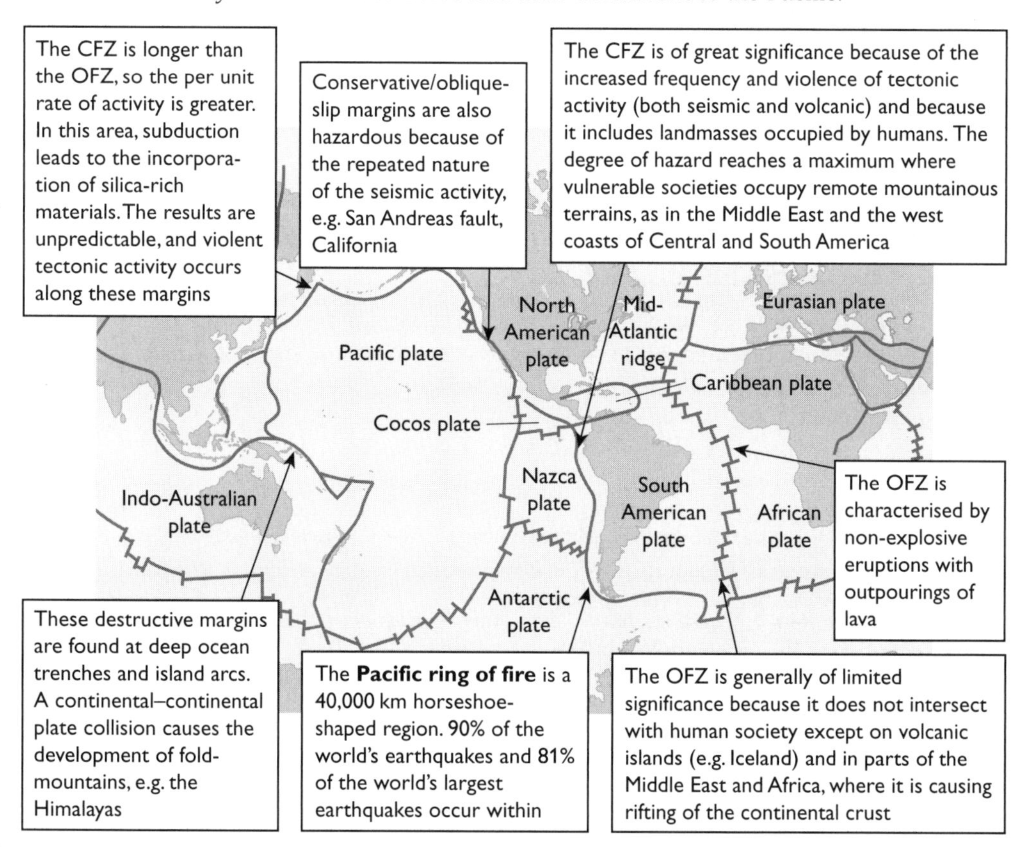

Figure 5 The geography of tectonic activity

Earthquakes and their associated volcanic activity at tectonic boundaries can sometimes be likened to a giant 'zip' where energy is gradually transferred along the plate boundary. The seismic 'unzipping' at this boundary can create a series of interrelated tectonic events, for example the Sunda fault off the coast of Sumatra where the Indian plate is being subducted beneath the Burma plate. Recently there have been two major earthquakes in Sumatra (2005) and Java (2006) with further seismic activity in 2007 and 2008.

There are two different types of crust: continental and oceanic (Table 3).

Table 3 The Earth's crust

	Oceanic	Continental
Depth (km)	5–9	30–70
Density (g cm^{-3})	3.0	2.7
Main rock type	Basalt	Granite

Research activity
How are the characteristics indicated in Table 3 important in determining the nature and impacts of both earthquake and volcanic activity?

Key reference: USGS website: **http://vulcan.wr.usgs.gov/Servers/earth_servers.html**

Enquiry question 2

What impact does tectonic activity have on landscapes and why does this impact vary?

Both earthquakes and volcanoes can have significant impacts on landscapes, and create specialist features or morphology. Volcanoes, for example, can create impressive volcanic cones (shield, acid cone) and lava plateaus (e.g. Deccan plateau in India). Generally less imposing are the intrusive volcanic features such as batholiths, sills and dykes, often associated with oozing of magma.

Research activity
Copy and complete the table to consider differences between landscape impacts, intrusive and extrusive igneous activity. Support each type with a located example, including facts and figures.

Extrusive igneous activity	Case study/ example/location	Intrusive igneous activity	Case study/ example/location

The impacts of earthquakes on landscapes are often less dramatic and diverse. At divergent plate margins (where two oceanic plates are diverging, such as the mid-Atlantic), **rift valleys** may form. They can also develop where an area of continental crust is being stretched by divergence, such as in eastern Africa. In a **normal fault** (again, common at divergent plate margins) a cliff-like feature known as a fault scarp may form. These can range from a few metres to hundreds of metres in height and their length may continue for 300 or more kilometres.

Research activity

Find out about divergent and convergent plate boundaries, normal and reverse faults, strike-slip faults (right and left). Also research rift valleys at divergent boundaries.

Key references for rock type and explosivity: **www.geology.sdsu.edu/how_volcanoes _work/**

Enquiry question 3

What impacts do tectonic hazards have on people and how do these impacts vary?

It is predicted that by 2025, 600 million people will be living in tectonically active areas. There are a number of reasons why people may live in tectonically dangerous or risky locations:

- People may stay in hazardous locations because of a lack of choice or alternatives. Very poor people may have to live in unsafe locations, close to volcanoes or in areas susceptible to earthquakes.
- Many people subconsciously weigh up benefits and costs, e.g. the economic benefits may outweigh the perceived risks of living on fertile farming land on the flanks of a volcano.
- Perceptions of hazard risks can be optimistic — having faith in technology to overcome/negate risks.
- There may be traditional, historical or cultural reasons for living close to volcanoes, e.g. people believing volcanoes were 'sleeping gods' — inertia has prevented people from moving away from such locations.

The number of deaths from volcanoes in recent years runs at about 1,000 per year, which is far greater than the number of deaths for previous centuries. However, the number of deaths from this type of tectonic hazard is much lower than that for earthquakes or hydro-meteorological events such as floods and droughts.

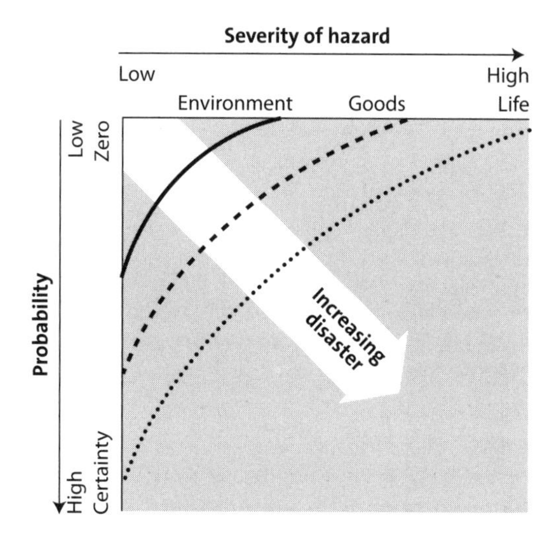

Figure 6 Relationship between severity of hazard, its probability and the degree of risk

Hazard types

There is a range of hazard types associated with tectonic events:

- earthquake, e.g. ground displacement, landslides, liquefaction, tsunamis, fires
- volcano, e.g. explosive blast, lava flows, ash flows and ash falls, mud flows, release of poisonous gases

Research activity

Research the above hazards (and other resulting hazards/secondary hazards, e.g. glacial outbursts caused by a volcano under a mass of ice) to develop a selection of case studies that can be used in the exam.

Key references for plate boundaries: **www.platetectonics.com/book/page_5.asp**
NASA Earth Observatory: **http://earthobservatory.nasa.gov/NaturalHazards/category.php?cat_id=12**

Some areas show concentrations of tectonic and other hazards. These are known as **multiple hazard zones**. For example, the Philippines in southwest Asia is at risk from a number of hazards, including tectonic and hydro-meteorological hazards (see Figure 7).

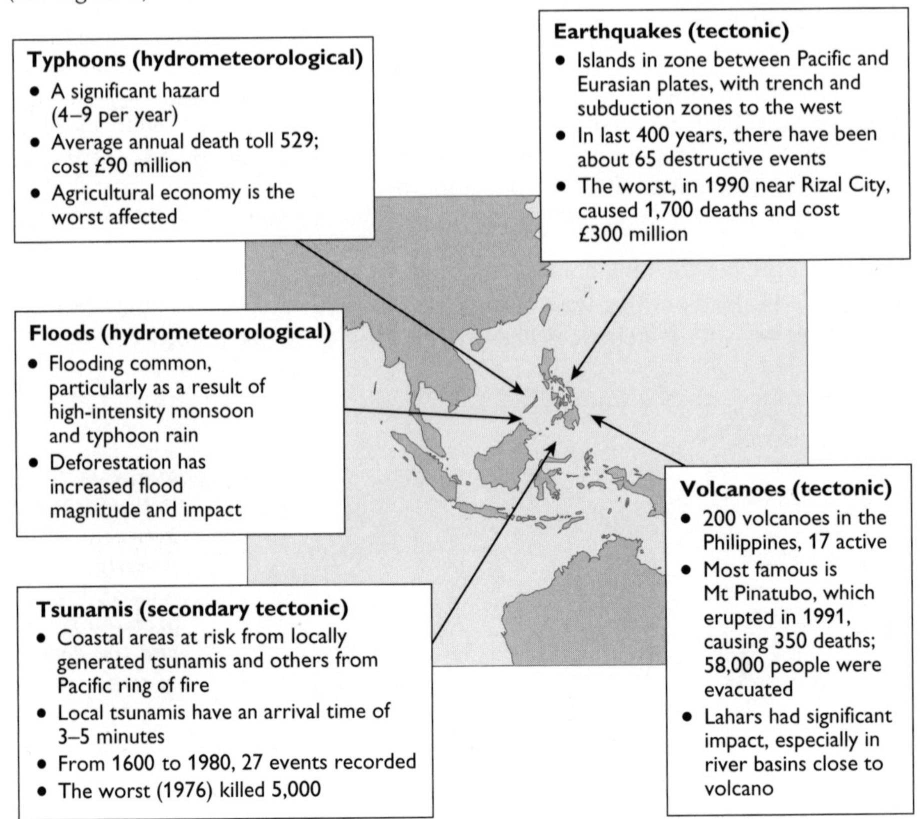

Figure 7 Hazards in the Philippines

Impact types

The impacts that tectonic hazards have on people can be grouped into social and economic (Table 4).

Table 4 Impacts of tectonic hazards

	Volcanoes and earthquakes — example impacts
Social	Human costs, e.g. primary, secondary and tertiary casualties. Hazards can also cause misery and suffering as well as poor health, e.g. loss of electricity/ services/infrastructure
Economic	Cost of repairing damage. Indirect costs/secondary impacts, such as loss of earnings, loss of tourism income

Note that tectonic hazards can have *beneficial impacts* on people. These include geothermal sources of energy, rich fertile soils and minerals such as gold. There are also potential income streams associated with tourism foreign exchange, e.g. visitors to Iceland.

Since 1950 the number of reported disasters for all hazard types (hydro-meteorological, biological and geological) shows an average rising trend. But for tectonic disasters (earthquakes, tsunamis and volcanoes) there is generally a fluctuating trend that does not show a noticeable decrease or increase. Some researchers have provided data to show a slight increase in volcanic activity. This remains controversial.

Trends in impact

Impacts can be measured in a number of ways, but typically they are reported in terms of number of deaths or casualties together with economic impact.

Figure 8 shows the number of deaths from geological hazards (including landslides/ mass movements). Impacts seem to have been relatively constant over time. Note that there are fewer deaths and casualties associated with tectonic hazards compared to other hazard types such as floods and droughts.

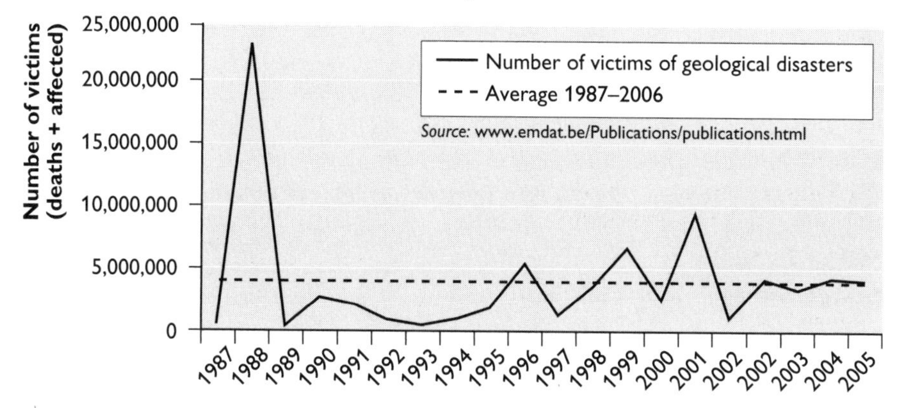

Figure 8 Deaths from geological hazards

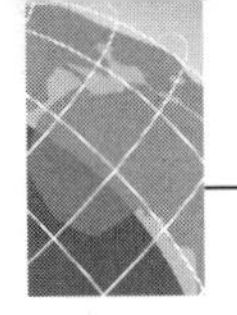

However, as a researcher, you should question the reliability of data — who provided them and how were they collected? Deaths from earthquakes and volcanoes in remote areas are typically underreported. Impacts need to be studied at different scales and in countries with different states of development.

Research activity
What evidence is there from the data that trends in impact have been influenced by anomalous years? Research additional data and statistics from the CRED website: **www.emdat.be/Publications/publications.html**

Enquiry question 4
How do people cope with tectonic hazards and what are the issues for the future?

Figure 9 shows how responses to tectonic hazards vary according to a number of factors, both physical and human.

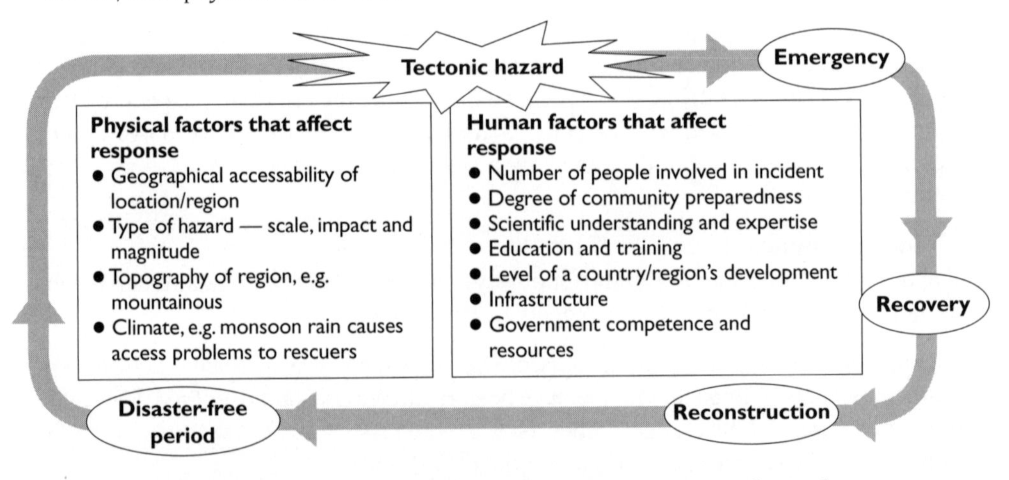

Figure 9 Factors that influence human response to hazards

Range of responses

The response to natural hazards can occur at a range of scales, from the individual within the local community to regional, national and international level. Various 'players' may be involved in hazard response (Table 5).

Table 5 The players involved in hazard response

National government	National policies, civil protection and defence, public information
Local government	Operational polices at the local level, e.g. rescue, welfare, medical, transport, supply of emergency aid
Scientists/academics/ educators	Researching, understanding causes, producing hazard (risk) maps. Also dissemination of information to public — raising awareness

Insurers	Risk assessment prior to hazard. Finance and assistance post hazard
Planners	Reducing risk by land-use zonation and planning, e.g. restricting development on floodplain
Relief agencies	Post-disaster aid, e.g. International Red Cross at 2004 tsunami
Engineers and architects	Design of buildings and infrastructures, e.g. Japanese technologists
Emergency practitioners and services	Police, medical, fire, traffic control — may also include army. Co-ordination is an important role
Media	In highlighting the tectonic hazard event (magnitude, scale, location, impact, duration etc.) Take on a warning role, especially important for more remote communities
Communities — bottom up	Important role in managing situations, as well as in terms of community preparedness and education

Specific strategies — before during and after the event

The choice of strategy will vary during the different phases of a hazard. Figure 10 is an attempt to model the impact of a disaster from pre- to post-disaster. It also considers the role of emergency relief and rehabilitation (see 'players' above).

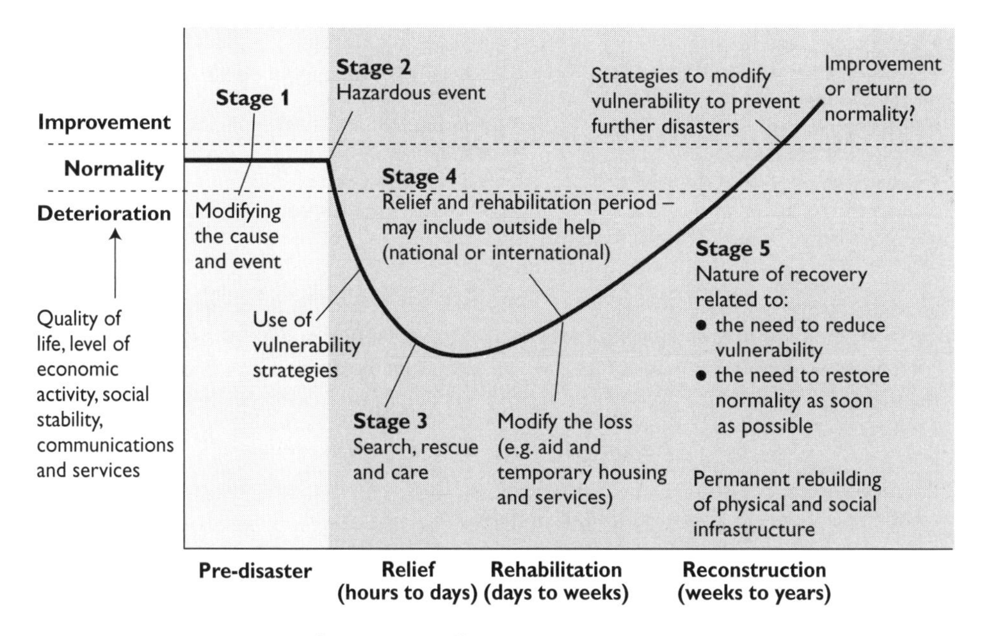

Figure 10 A disaster-response curve

Figure 11 gives a useful framework for response analysis. Approaches to managing the tectonic hazards vary by time as well as place.

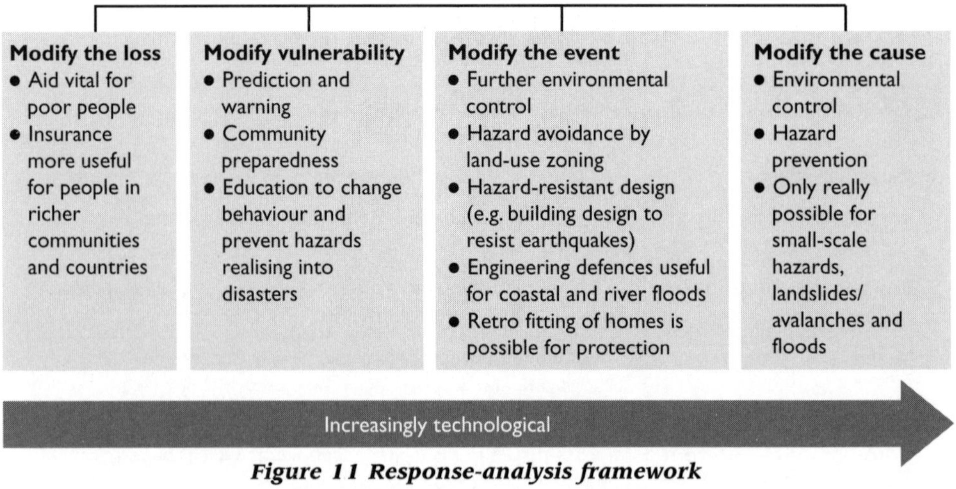

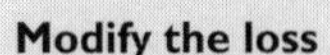

Modify the loss	Modify vulnerability	Modify the event	Modify the cause
• Aid vital for poor people • Insurance more useful for people in richer communities and countries	• Prediction and warning • Community preparedness • Education to change behaviour and prevent hazards realising into disasters	• Further environmental control • Hazard avoidance by land-use zoning • Hazard-resistant design (e.g. building design to resist earthquakes) • Engineering defences useful for coastal and river floods • Retro fitting of homes is possible for protection	• Environmental control • Hazard prevention • Only really possible for small-scale hazards, landslides/ avalanches and floods

Increasingly technological →

Figure 11 Response-analysis framework

There is also an interesting balance that needs consideration between increasing vulnerability and risk — see Table 6.

Table 6 The balance between vulnerability and risk

Increased vulnerability	Reduced vulnerability
• Population growth • Urbanisation and urban sprawl • Environmental degradation • Loss of community memory about hazards • Ageing population • Ageing infrastructure • Greater reliance on power, water, communication systems • Over-reliance on technological fix	• Warning and emergency-response systems • Economic wealth • Government disaster-assistance programmes • Insurance • Community initiatives • Scientific understanding • Hazard engineering

Research activity

Copy and complete the table below (earthquakes vs volcanoes) to give an analysis of types of responses to the management of tectonic hazards. You could also support each comment with a located example.

	Advantage	Disadvantage
Land-use control/design resistant buildings	Can save lives; can be easily regulated through planning controls etc. ...	
Education and training		
GIS/remote sensing		Cost (technology gap — MEDC vs LEDC); reliability; coverage...

Cold environments: landscapes and change

Introduction

Cold environments occur on a large percentage of the Earth's surface. There are also many other locations that show evidence of having been cold environments in the past as a result of the historically cooler climatic conditions that have existed in parts of the world through geological time.

These cold environments possess a variety of landform types and features, reflecting the diverse nature of the physical and human processes that have helped shape them. These processes are greatly influenced by current and past climatic conditions, especially by shorter and longer-term variations in temperature.

Cold environments can provide many opportunities for human activity and as a result there can be conflicts in terms of management, especially exploitation vs protection. Different attitudes exist within the various interested parties and this can lead to significant conflicts between them.

You will need to research a range of glacial and periglacial environments, both present and relic. In particular, you will need to find out about the distribution of different types of cold environments.

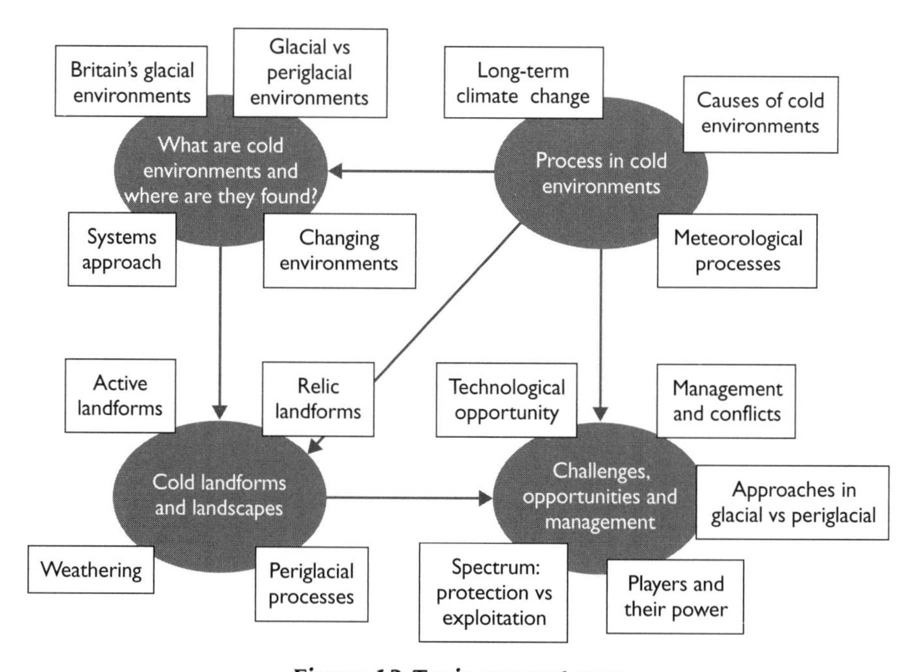

Figure 12 Topic concept map

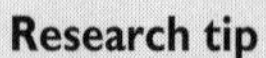

Research tip
When you research case studies and smaller examples make notes on all the enquiry questions, with a log of sources in case you need to add details on one area once the research focus pre-release material is available.

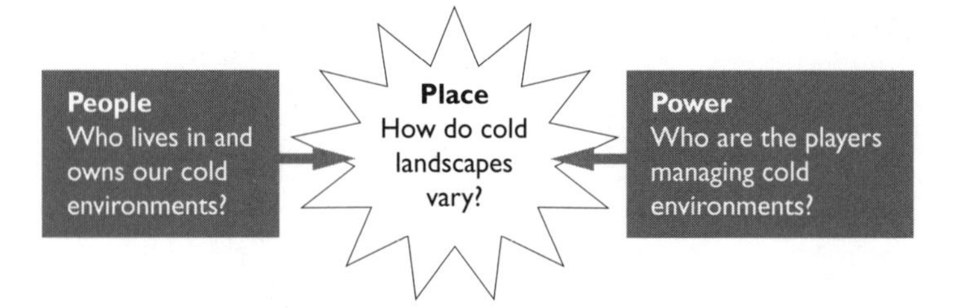

Figure 13 The synoptic context

Table 7 Checklist for cold landscapes

Enquiry question 1 **What are cold environments and where are they found?**	Enquiry question 2 **What are the climatic processes that cause cold environments and what environmental conditions result from these?**	Enquiry question 3 **How do geomorphological processes produce distinctive landscapes and landforms in cold environments?**	Enquiry question 4 **What challenges and opportunities exist in cold environments and what management issues might result from their use?**
• What are the different types of cold environments? • Glacial vs periglacial • Global patterns and distribution • Short vs long-term changes in distribution • Glacials vs interglacials (especially history in British Isles) • Landscape systems • What is mass balance?	• Short vs long-term changes in climate — atmospheric 'triggers' • Global systems as distributing heat energy • Where were past glacial and periglacial environments? • Climate change and future for cold environments	• How different are ice sheet and valley glacier landscapes? • How different are upland and lowland ice landscapes? • Scale and energy of impacts • Relic vs active ice landscapes • Periglacial landscapes and landforms	• What are the main challenges in living in/exploiting cold environments? • What opportunities do cold environments present? • How and why are there different attitudes to protection or exploitation of cold landscape resources?
Key refs used	Key refs used	Key refs used	Key refs used

Research activity
Use Table 7 as an initial stimulus, then carry out your own research audit to obtain background sources of information. Make a note of any key references used.

Enquiry question 1

What are cold environments and where are they found?

Cold environments are icy landscapes that occur in (a) polar areas and (b) high mountainous regions of the world. They are characterised by very cold temperatures together with lots of snow and ice, and few plants and animals. They have also experienced relatively little commercial exploitation and have low human population densities. But not all 'cold' environments are the same and they can be subdivided into: (a) glacial areas, (b) tundra and (c) alpine areas or high mountain ranges.

Understanding the basic concepts

Many cold environments can be understood by way of a systems approach, i.e. by thinking about three distinct elements: *inputs*, *processes/transfers* and *outputs* (Figure 14).

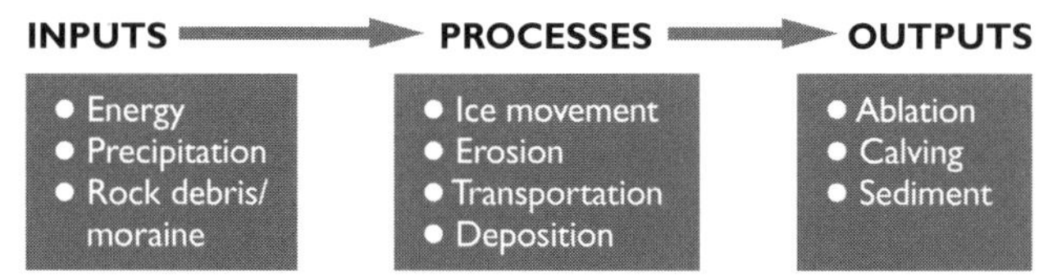

Figure 14 Inputs, processes and outputs

The balance between the inputs and outputs of snow and ice in a glacier (Figure 15) is called the **mass balance**. This can be either positive or negative (Figure 16).

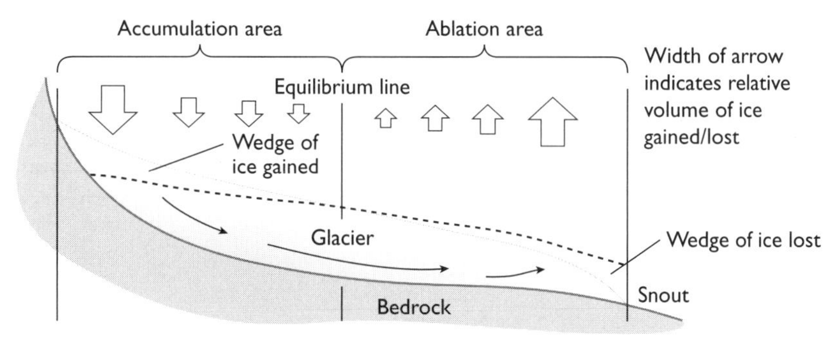

Figure 15 The glacial system

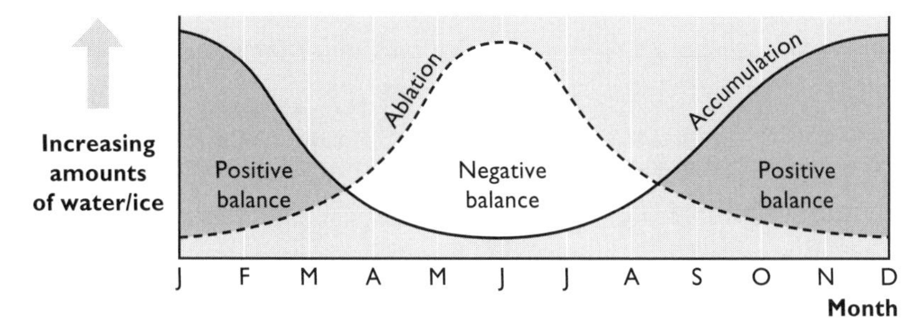

Figure 16 The net budget in a northern hemisphere glacier

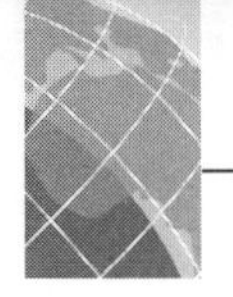

There will often be seasonal variations in the budget with accumulation exceeding ablation in the 'winter' and vice versa in the 'summer'. It is possible that there will be an advance during the year even if the net budget is negative and some retreat even when it is positive. Even in retreat, a glacier may advance under gravity in front of the **firn line**, so it can appear that a retreating glacier is actually advancing.

Some other important terminology is summarised in Table 8.

Table 8 Glacial terminology

Frequency and magnitude	Frequency refers to how often an event happens or repeats itself. Magnitude refers to the size of an event. In glacial systems high frequency/low magnitude events include the daily abrasion carried out beneath a glacier. Some events happen very infrequently (low frequency) but can be catastrophic in terms of power and impact on the landscape, e.g. a glacier burst (jökulhaulp).
Equifinality	The notion of equifinality states that in an open system the same end result can be reached by many different means. In glacial environments this means that different processes may produce the same landform or landscape feature. For instance, outwash plains may be produced by a small number of major glacial bursts, as well as by small, incremental amounts of fluvioglacial deposition over a longer period of time.
Dynamic equilibrium	Natural systems have a tendency towards achieving a steady balance. If something happens to disturb the equilibrium of a balanced system, the system itself produces a dynamic response in order that equilibrium can be achieved. But equilibriums can change over time and under different conditions. For example, in glaciers an increase in accumulation leads to a growth in the glacier. This in turn advances the equilibrium/firn line until the system returns to a stable state.

Changing cold environments

The distribution of cold environments has changed throughout geological and more recent history. In the past, cold landscapes have been more extensive than at present (Table 9) (nearly all the ice on Earth today is locked up in Greenland and Antarctica). One controversial theory, 'Snowball Earth' suggests that at one point in history the whole planet was covered in ice. Even if this contentious idea is not true, what is certain is that there have been repeated glacial periods (glacials) separated by warmer interglacial times (Figure 17). The last glacial maximum occurred about 20,000 years ago — this was the latter part of the Pleistocene ice age.

Table 9 Ice coverage in 1970 compared to maximum for selected world regions

Region	Area of ice in 1970 (km^2)	Maximum area during glacial maximum (km^2) — estimated
Antarctica	12.5 million	13.8 million
Greenland	1.7 million	2.3 million
Asia	115,000	3.9 million
South America	26,000	870,000
Europe	3,600	37,000

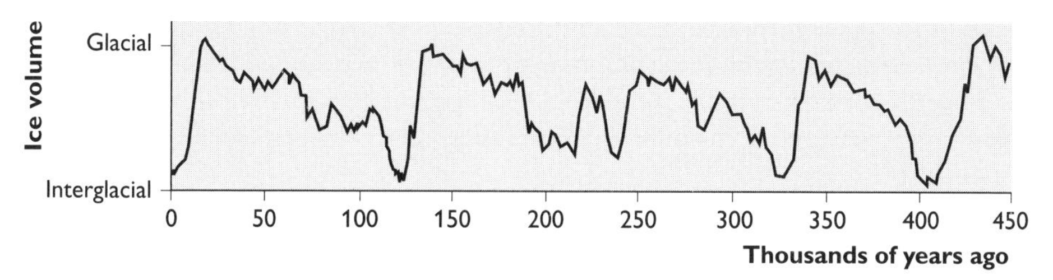

Figure 17 Glacial and interglacial periods of the last 0.5 million years

Oxygen-isotope ratios can be used to reconstruct past climates. Note the fluctuations during the Devensian period in Figure 18 — a series of glacials and interglacials.

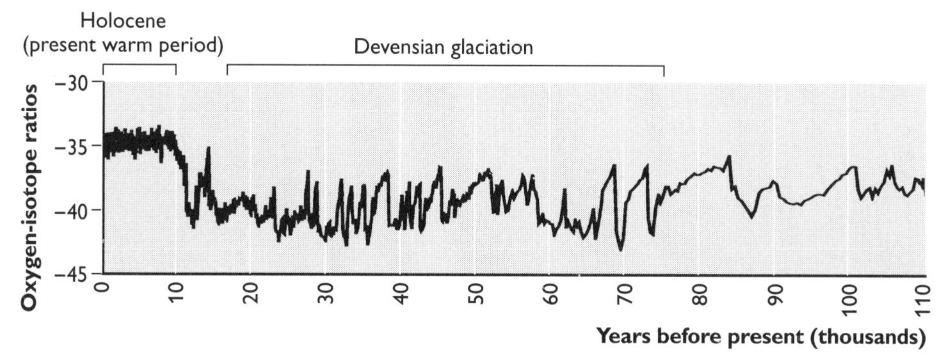

Figure 18 Oxygen-isotope ratios

Past ice in Britain

In the past, when the climate was much colder, there were several significant glacial environments in Britain. They were found mainly in Scotland, the Lake District and North Wales as well as in parts of Ireland. Britain experienced several glacial periods during the last 2 million years of the Quaternary period (Figure 19).

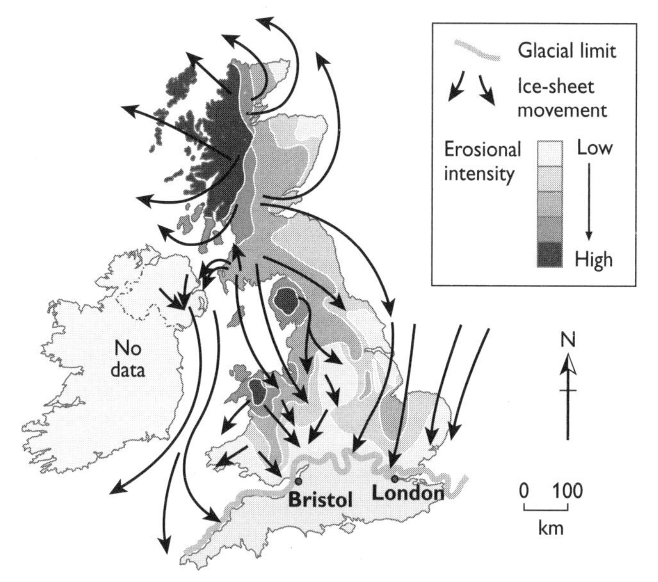

Figure 19 The glacial maximum in Britain

Research activity
Find out more about the ice age in one part of Britain or overseas. Key reference for impact and evidence in north Norfolk: **www.bbc.co.uk/norfolk/content/articles/2005/06/21/coast05_iceage_feature.shtml**

Enquiry question 2

What are the climatic processes that cause cold environments and what environmental conditions result from these?

Factors affecting temperature

The low temperatures that characterise cold climate environments are the result of four factors:

1. **Latitude** — in high latitude areas (i.e. 60°–90°N and S of the equator) the sun in the sky is low. Incoming solar radiation passes through a greater thickness of atmosphere compared to mid-latitude locations (see Figure 20). Energy is lost due to absorption, scattering and reflection (e.g. from surfaces such as cloud tops).

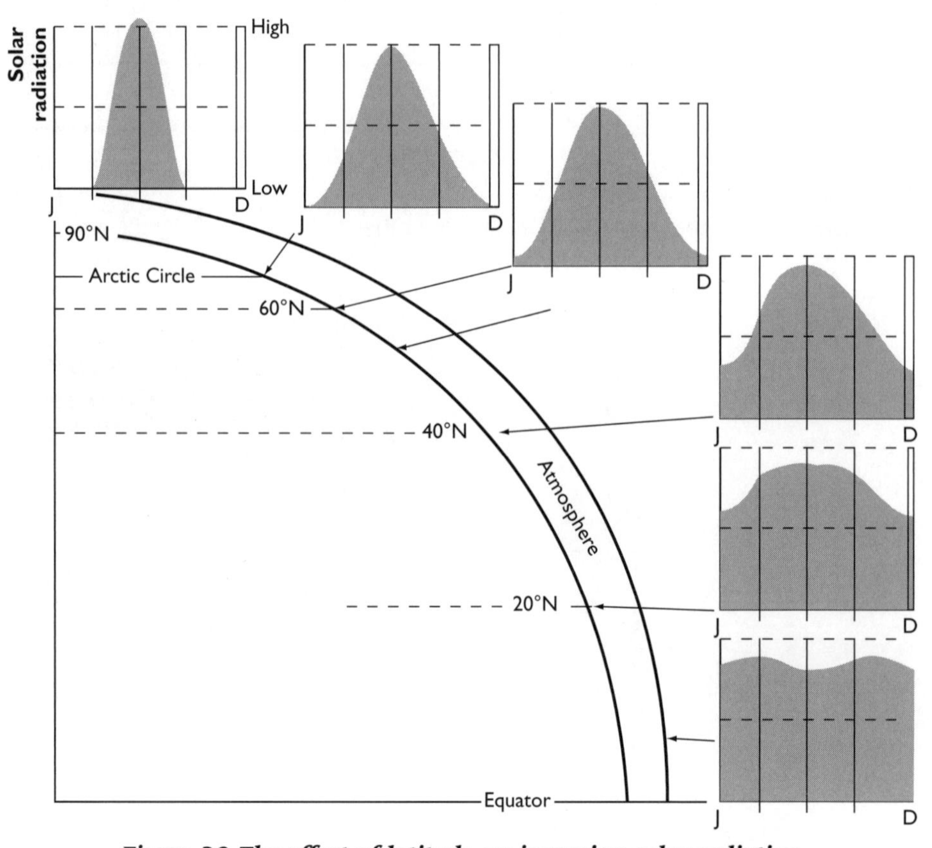

Figure 20 The effect of latitude on incoming solar radiation

Also, the low angle of incidence means that the remaining solar energy is spread over a wider area, and much of this incident energy is reflected back into the atmosphere from the surface of snow and ice (high albedo or reflectivity).

The other implication of latitude is the resultant pressure systems. Cold, dense air tends to sink at or near the poles and produces high-pressure polar anticyclones. Air from these pressure systems moves slowly outwards, spreading into the high mid-latitude areas causing low temperatures.

2 **Altitude** — temperatures decline with altitude because air becomes thinner with height above sea level and is less able to absorb heat from outgoing, long-wave terrestrial radiation. The decrease in temperature with altitude is called the environmental lapse rate and on average is about 6°C per 1,000 m. This fall in temperature explains why glaciers can exist near the equator, but only at altitudes above 5,000 m, e.g. Cotopaxi and Cayambe volcanoes in Ecuador.

3 **Continentality/distribution of land and sea** — the ability of land and sea to absorb heat and radiate it back into the air above varies greatly. In general, land (soil and rock) warms up more quickly, and is able to release the stored heat at a faster rate compared to water. This means continental interiors are warmer than coastal areas in the summer, but much colder than coastal areas during winter because the land has lost its stored heat, whereas the sea is still radiating heat that it absorbed during the previous summer.

4 **Ocean currents** — The importance of ocean circulation has only been revealed relatively recently. It has now been demonstrated how the warm North Atlantic Drift (NAD), for example, partly controls temperatures in Western Europe. The current transfers heat energy from the Gulf of Mexico northwards across the Atlantic (Figure 21).

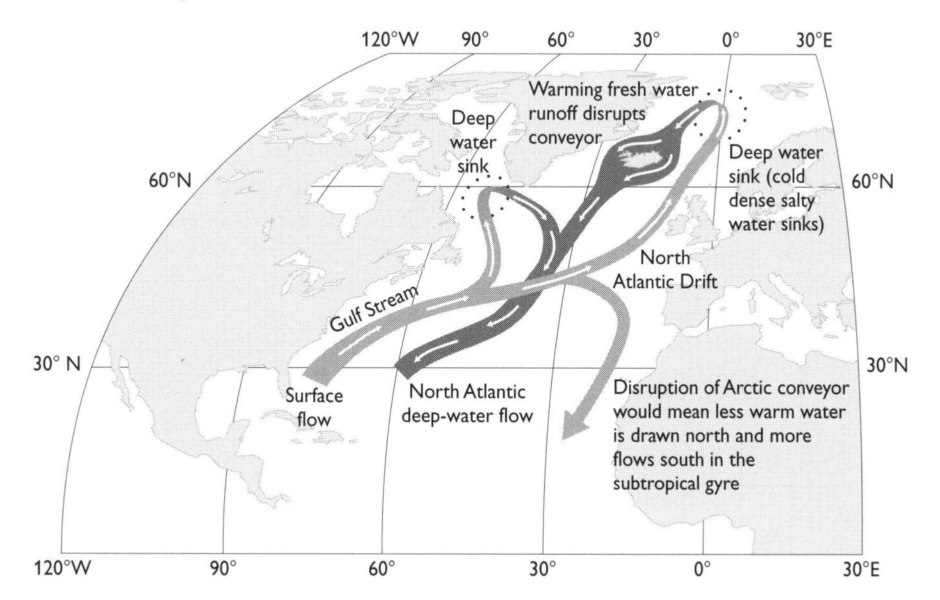

Figure 21 The path of the North Atlantic Drift oceanic current

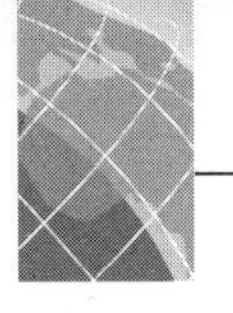

There are concerns that large influxes of fresh water (e.g. from the melting of ice sheets or greater discharge of rivers) can actually turn off this heat conveyor belt. A loss of the NAD would have significant impacts on Western Europe both in terms of environments and economies.

Longer-term climate change

Throughout the Earth's history temperatures on the surface of the planet have varied considerably (see Figure 17). Many different causes of these medium and longer term climatic changes have been proposed and include changes in the Earth's orbit, the tilt of the Earth's axis, solar radiation, position of the poles, the amount of water vapour in the atmosphere, distribution of land and sea, flows of ocean currents and the amount of carbon dioxide in the atmosphere.

Research activity
Find out more about the longer term causes of climate change. Which are the most significant? See the British Antarctic Survey for more details: **www.antarctica.ac.uk/bas_research/science/climate/climate_change.php**

Enquiry question 3

How do geomorphological processes produce distinctive landscapes and landforms in cold environments?

Geomorphology is the study of landforms, landscapes and the processes that create them. Areas that have been glaciated usually have distinctive landscapes because ice has significant power to modify existing landforms. A well-used example is where ice in a valley glacier changes the scale and shape of a pre-existing river valley (from v-shaped to more of a trough shape) — see Figure 23.

Processes of erosion

Two main erosional processes, abrasion and plucking, are responsible for some of the most impressive erosional landforms on the planet:

- **Abrasion** occurs where glacier ice is in contact with the valley bottom and sides. Ice at the bottom of the glacier carries debris and scratches or abrades the bedrock surface. Figure 22 shows the factors that can affect abrasion. The process of abrasion is evident in the form of striations — scratches on bedrock that indicate the direction of ice flow.

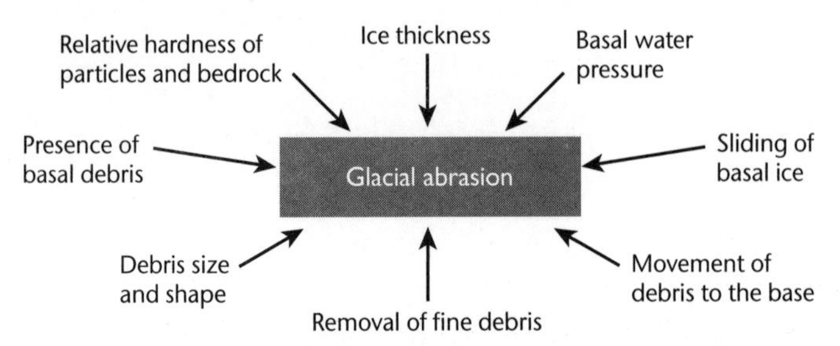

Figure 22 The factors that can affect the amount and rate of glacial abrasion

- **Plucking**, or 'joint block removal' is also dependent on the movement of ice. Ice freezes to jointed bedrock and then the rock becomes incorporated into the sole of the glacier as it moves. This is only possible due to basal pressure melting whereby the base of the glacier can thaw and re-freeze (re-gelation) due to changes in the pressure exerted by the overlying ice.

Rates of glacial erosion will vary enormously in both space and time. In 1968, two researchers calculated that mean annual erosion for active glaciers is between 1,000 and 5,000 m^3. In comparison, ice 100 m thick flowing at 250 m per year in the French Alps managed to lower a marble plate by 36 mm in one year.

Research activity
Find out more about the how the factors shown in Figure 22 will affect the amount and rate of abrasion carried out by a glacier. Also research examples of different glaciers to illustrate these various factors. Visit the World Glacier Inventory: **http://nsidc.org/data/g01130.html**

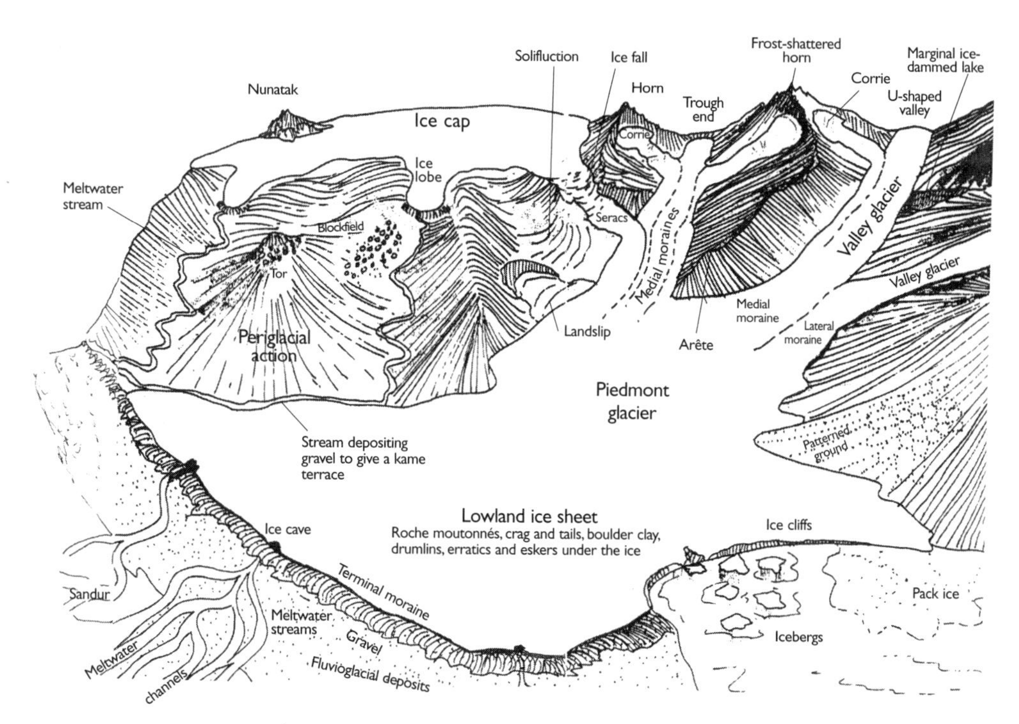

Figure 23 Range of glacial landform types

Sub-aerial processes may improve the efficiency and impact of abrasion and plucking by providing additional weathered rock debris:

- **Freeze–thaw** — the pressure exerted by water changing state and expanding into ice. This process, when repeated, can be highly effective at the destruction of heavily jointed rock, particularly material exposed above the level of ice.

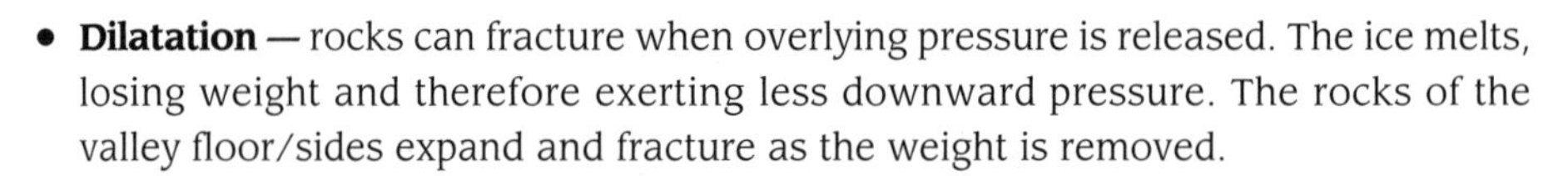

- **Dilatation** — rocks can fracture when overlying pressure is released. The ice melts, losing weight and therefore exerting less downward pressure. The rocks of the valley floor/sides expand and fracture as the weight is removed.

Distinctive landforms

An important task for this module is to research both upland and lowland landforms of glaciation. (Figure 23 on p. 35 shows an example of the range of landform types.) For each landscape feature (erosional, depositional, periglacial, upland, lowland etc.) you will need to know about:

- processes which formed the feature (including the role of weathering)
- scale and distribution (within the glacial system and regionally), plus characteristic shape/patterns
- examples: location, size, special characteristics etc.

You will also need to research **periglacial environments** — both above ground and below ground processes and features. Table 10 provides a list of periglacial features and locations that can act as a starting point.

Table 10 Periglacial environments and example locations

Feature	Examples
Ice wedge casts	Giant ice wedge casts at Stanton Harcourt and Baston in Lincolnshire
Patterned ground	Stripes in Lincolnshire and Yorkshire wolds; nets and stripes on Thetford Heath, East Anglia; centre of nets correspond to chalk-loving grasses, while the 'net' reflects heathers growing on thick sands
Cryoturbation structures (deformation of sediments near surface, for example frost heave)	Isle of Thanet, Kent; East Anglian chalklands
Pingos	Walton Common, Cambridgeshire, where 'scars' are superimposed, indicating that several cycles of ground-ice growth are represented; many rampart repressions in valleys of southwest Wales, e.g. Cletwr Valley
Thermokarst scenery-slumps and depressions caused by melting ground-ice	Walton Common, Grunty Fenn, Cambridgeshire
Head deposits formed through frost shattering, solifluction and sliding	Sarsen stones — streams of surface boulders running along the valley bottom of the Marlborough Downs, e.g. Coombe rock, consisting of chalk, flint and mud
Asymmetric valleys (periglacial processes acting unequally on slopes as a result of their aspect)	Chilterns

Feature	Examples
Nivation terraces (steps cut into the slope profile caused by prolonged nivation) — frost-action, mass wasting and meltwater erosion beneath melting snowdrifts	Slopes of Cox Tor, Dartmoor
Tors	Penine tors (gritstone); Stiperstones, Shropshire (quartz site); Dartmoor tors (granite)

Enquiry question 4

What challenges and opportunities exist in cold environments and what management issues might result from their use?

Cold landscapes offer a range of threats, challenges and opportunities to people. Although only a small proportion of the world's population live in cold environments, many of us are either directly or indirectly influenced by cold locations. Periglacial and permafrost environments present their own challenges and opportunities.

Research activity

Find out more about the wide range of challenges (these can include hazards) that exist in cold environments. Put your findings into a table, similar to the example provided.

Threat	Description	Located example
Ice avalanche	Ice 'calved' from the snout in steep areas of relief. Sometimes ice avalanches occur together with landslides and snow avalanches...	Mt Iliamna, Alaska, in 1980 produced 20 m^3 of ice. Speeds of up to 100 km h^{-1}...
Flood	Catastrophic outburst floods (jökulhlaups). Meltwater is suddenly released...	...
Mud flow and lahar	...	...

Other challenges may be less obvious, but still difficult for people who live in such environments, e.g. problems of low agricultural productivity and difficulty in establishing and maintaining transport infrastructure.

Research activity
Over a period of 4–6 weeks, scan the international newspapers and their corresponding websites to identify any news items linked to cold environment challenges, e.g. management of fragile ecosystems, sea level rise (melting ice NOT thermal expansion), avalanches, floods, etc. Make brief notes on each so that they can be used as case studies in preparation for the exam. Key reference — newspapers of the world can be accessed at:
www.newseum.org/todaysfrontpages/flash/

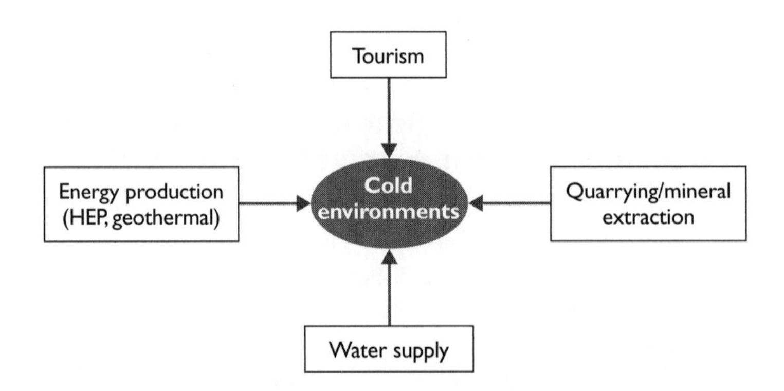

Figure 24 Cold environments also present many opportunities

Since the mid-nineteenth century, glaciers have provided the basis for particular types of tourism and now form the cornerstone of many local economies. For example, the tourist industries of Greenland and Iceland are based largely on their spectacular glacial landscapes. Some National Parks are based on present day glaciers, e.g. Glacier National Park in Montana, or on relic glacial landscapes, e.g. the Lake District in the UK.

Research activity
Investigate how technology can be used in both periglacial and glacial environments to overcome challenges linked to one element of exploitation, e.g. building a new HEP station, maintenance of fragile ecosystems.

Life on the margins: the food supply problem

Introduction

There is enough food in the world for everyone; the problem is its uneven distribution. Over 50% of the world's population live in low-income, food-deficit countries that are unable to produce or import enough food to feed their people. Most of the world's hungry people are in the developing world, but another 34 million live in the developed world. Life is on the margin of survival for these people in both rural and urban areas. Over one third of all children are malnourished and 6 million children a year die of causes related to malnutrition. One leading player in the food supply problem, the UN Food and Agriculture Organization (FAO), organises World Food Day on 16 October each year, to highlight its aim of food security: 'when all people have physical and economic access at all times to enough nutritious, safe food to lead healthy and active lives.'

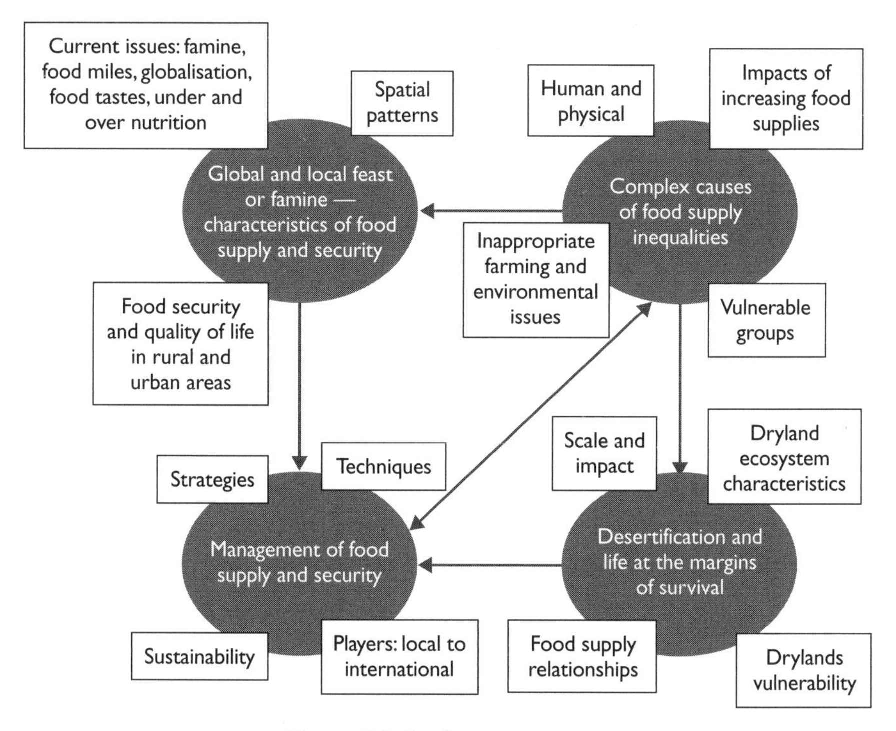

Figure 25 Topic concept map

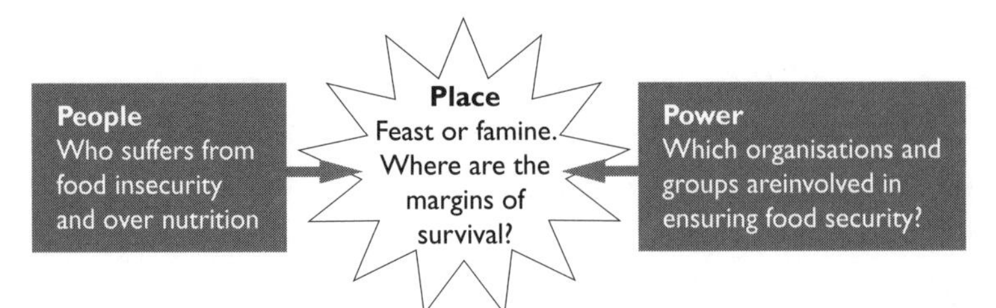

Figure 26 The synoptic context

Research activity

While thinking of the synoptic aspects of the course, a simple checklist based on the topic/enquiry questions from the full specification will help you separate your research into manageable chunks (see Table 11).

Key references: overviews, case studies, quizzes: **http://anthro.palomar.edu/change/default.htm**

2008 World Food Summit: **http://news.bbc.co.uk/1/hi/in_depth/7432583.stm** and **www.fao.org/newsroom/en/focus/2008/1000829/index.html**

'The silent tsunami', Food crisis of 2008: **www.mssrf.org/fs/index.htm** and **www.economist.com/world/international/displaystory.cfm?story_id=11049284**

Table 11 Checklist for life on the margins

Enquiry question 1 **What are the characteristics of food supply and security?**	**Enquiry question 2** **What has caused global inequalities in food supply and security?**	**Enquiry question 3** **What is the role of desertification in threatening life at the margins?**	**Enquiry question 4** **How effective can management strategies be in sustaining life at the margins?**
• What type of socioeconomic current issue? • Is it a global or more localised issue? • Any environmental issues? • Why does food supply vary spatially? • How does food security affect quality of life?	• Classify the causes into economic, social, environmental plus direct and indirect: long and short term • Is population pressure involved? • Is type of food production technique involved, and effects on environment? • Identify which groups of people most affected by food insecurity • Why are they affected?	• Scale of problem • Physical climatic causes? • Human causes? • Role of increased food supply • Is salinisation a problem? • Why are dryland ecosystems so vulnerable? • How is food supply affected by desertification?	• Who is involved in management? • Role of organisations/ groups involved from local to global • Types of strategies /techniques used (e.g. trade, aid, increased food production by high and low tech) • Reasons for those • How effective are the organisations/groups and strategies/ techniques? • Role of sustainability
Key refs used	Key refs used	Key refs used	Key refs used

Enquiry question 1

What are the characteristics of food supply and security?

Current issues in 'Life on the margins'

- Food insecurity — the unreliability of quality food supply can lead to famine and ultimately death (see Figure 27).

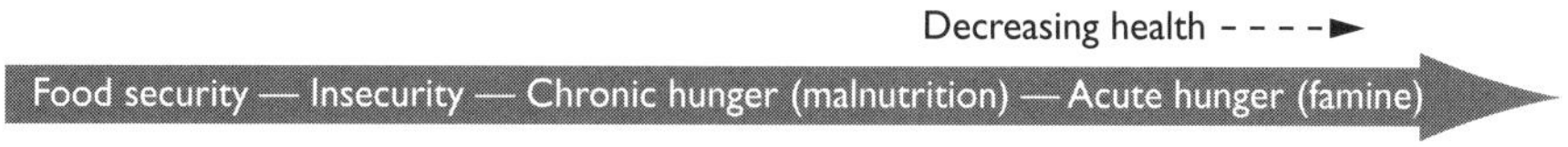

Decreasing health - - - -►

Food security — Insecurity — Chronic hunger (malnutrition) — Acute hunger (famine)

Figure 27 Food insecurity

- The areas most at risk are often arid and semi-arid areas such as the Sahel, but vulnerable groups are growing as the development gap widens, from Mumbai to New York, Bolivia to Cambodia — with a concentration in sub-Saharan Africa (see Figure 29).
- Food transition issues: overnutrition and obesity. This involves the globalisation of food tastes and the increased consumption of high protein, fats and sugars by rising middle classes especially in urban areas. Globally, obesity has now reached epidemic proportions.
- Food suppliers: the rise in power and influence of mega-supermarket chains and TNCs plays a large role in undercutting local food supply chains and producing cheap fast-food alternatives, especially for urban populations.
- Environmental issues: the negative effects of the scale and intensity of food production. Since the 1990s, issues have included the Green Revolution, GM crops, deforestation and growing biofuels in place of staple food crops.
- Food miles and footprints: the distance food travels from producer to consumer and environmental impacts of food supply.
- Food supply and prices crisis in 2008: causing global food scarcity.

What is food security?

Figure 28 shows that food must meet physiological requirements in terms of quantity, quality and safety and must be socially and culturally acceptable. Food security is determined at local and national scales by stability in availability, access, and use/utilisation. Where the nutritional status is low, there is no food security even though there might be a surplus of food supply at the national or regional level. It is clear that it will be difficult for managers to solve problems of insecurity, which often have complex causes.

There is a difference between:

- Transitory food insecurity — which may be either cyclical/seasonal (e.g. gaps in staple food production like rice during a 'lean season') or temporary (from unpredictable climate shocks like drought and floods).
- Chronic food insecurity — where vulnerable groups are permanently unable to ensure their food needs, usually linked to poverty.

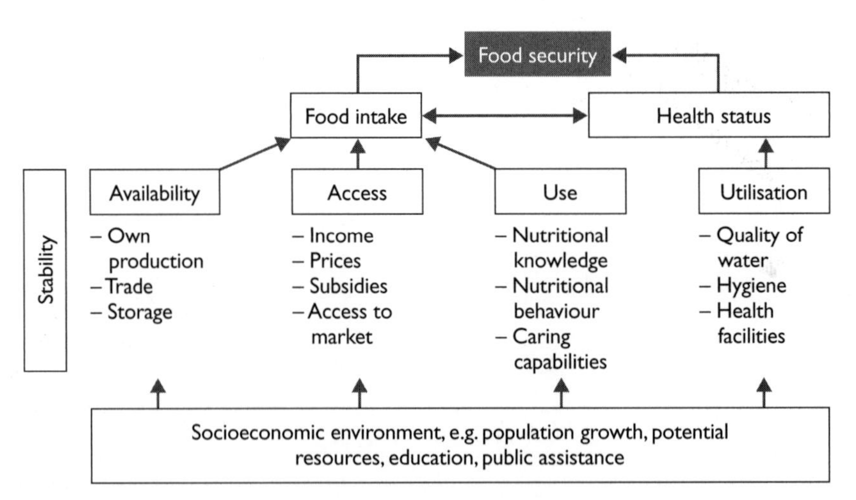

Figure 28 Food security

To be food secure means that:

- Food is available — the amount and quality of food available globally, nationally and locally can be affected temporarily or for long periods by many factors including climate, disasters, war, civil unrest, population size and growth, agricultural practices, environment, social status and trade.
- Food is affordable — when there is a shortage, food prices increase, and while richer people will likely still be able to feed themselves, poorer people may have difficulty obtaining sufficient safe and nutritious food without assistance.
- Food is utilised — at the household level, sufficient and varied food needs to be prepared safely so that people can grow and develop normally, meet their energy needs and avoid disease.

The physiological effects of food insecurity include:

- Intergenerational problems — low calorie and nutrients intake slows metabolism, reducing ability to work and makes food more difficult to utilise. A hungry mother will have underweight babies who will have stunted growth, frequent illness, learning disabilities and reduced resistance to disease.
- Contaminated food and water can also cause illness, nutrient loss and often death in children.
- Acute malnutrition is the result of sudden weight loss due to starvation and disease, often leading to rapid death as it increases the risk of infection and vital organs stop working. It is treatable but it may have long-term effects on physical and mental growth.
- 'Wasting' means children are far thinner for their height than healthy children.

Issues resulting from increased food production methods

Problems resulting from increased food production of common UK food types (e.g. orange juice, cooking oil, milk, chicken, beef, fish, sugar, fruit and vegetables) include:

- pollution issues, e.g. waste, chemicals, emissions
- loss of biodiversity

- ecofootprint
- food miles
- ethical issues, e.g. animal welfare, social issues (exploitation of workers)

Why food supply varies spatially

Some areas are naturally more suited to food production, e.g. the great plains of America and Russia. Urban fringes have also been traditionally important in food supply, but are under constant threat from sprawling urban growth.

Physical factors limit food production unless technology is available to overcome temperature, water and nutrient deficiencies. Irrigation, chemicals and greenhouses are, however, costly. Agricultural advances in yields have shown there is a law of diminishing returns that limits even GM products as a supply source. Table 12 lists human and physical reasons for variation in food supply.

Table 12 Factors that influence food supply variations from place to place

Human factors	Physical factors
Accessibility of markets Land ownership systems — rented land or that without secure tenure causes insecurity Inheritance laws — may be gender biased Market and trade patterns and regulations skewed in favour of more developed economies Competition, which can be healthy but often unfair if subsidies, quotas, etc. involved Government action and support Role of businesses and TNCs, which now dominate research globally and are governed by profit margins, not necessarily food security of the poorest Role of aid agencies in both long- and short-term food supplies	Soil — nutrient store Climate — seasonal changes Precipitation — amount, frequency, type Length of thermal growing season Relief — steep or waterlogged areas less useful Aspect — slope angle Altitude — affecting temperature, water supply Hazards — tectonic, hydro-meteorological and biological. Northern China produces 58% of the country's food crops, is suffering its worst drought in 50 years and highlights the increasing concern over climate change reducing agricultural production

Research activity

Contrast and compare two key case studies on food intensification, security and resulting socioeconomic and environmental problems: the EU's Common Agricultural Policy and the Green Revolution based in Asia and Latin America. Create a timeline for each to show changes, and why the new Green Revolution of the twenty-first century, targeting Africa, may make food security more sustainable.

Key references: short podcasts of past and present Green Revolutions, food transitions and impacts on environment: **http://news.bbc.co.uk/1/hi/programmes/documentary_archive/6500041.stm**

Greening of the Green Revolution: **www.nature.com/nature/journal/v396/n6708/full/396211a0.html**

Africa's food crisis:
http://news.bbc.co.uk/1/hi/in_depth/africa/2006/africa_food_crisis/default.stm

Food security in urban and rural areas

Traditionally food insecurity is a characteristic of rural areas, especially in poorer economies. Figure 29 shows that sub-Saharan Africa has the highest rate of under-nourishment.

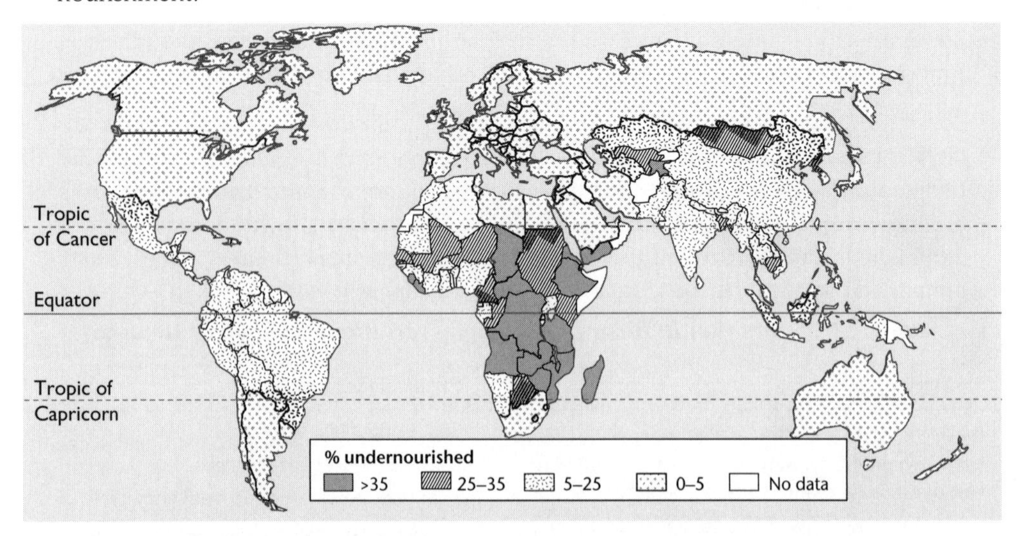

Figure 29 Global distribution of under-nourishment, 2003–05

India still ranks ninety-fourth in the UN World Food Programme Global Hunger Index of 119 countries. Large-scale famines have been controlled since independence in 1947, the Green Revolution has boosted food production, especially in the Punjab. However, according to the *Food Insecurity Atlas of India* produced by the Swaminathan Research Foundation, it now has the highest absolute number of undernourished people in the world, especially in small towns.

Urban food insecurity is not confined to developing nations: the New York City Coalition Against Hunger estimated that by 2006 1.3 million New York residents, including 400,000 children, lived in households with an insecure food supply. Michelle Obama made food security a headline in 2009 by helping in a Washington soup kitchen: food insecurity is on the increase globally. People on low incomes, wherever they are, will almost inevitably be more vulnerable to food security issues than those who are more affluent.

Enquiry question 2

What has caused global inequalities in food supply and security?

Food production and food security

Food production varies greatly. Between 1960 and 1990:

- World cereal production more than doubled.
- Food production increased by one third per head.

- Daily intake of calories increased by one-third.
- Real food prices fell by almost half.

Yet, 963 million people, more than the populations of the USA, Canada and the EU, do not have enough to eat — 907 million of these are in developing countries. The FAO estimates that every 6 seconds a child dies, and the number of undernourished people in the world increased by 75 million in 2007, largely due to higher food prices.

Soil degradation, chronic water shortages, inappropriate agricultural policies and population growth threaten food production in many countries. While growing export crops such as coffee, cocoa and sugar produces export income, it can lead to a decrease in basic food production, causing hardship for people who are poor. These complex causes may be categorised into root and direct causes (Table 13).

Table 13 Factors affecting food security

	Direct causes	Root causes
Economic	Income, poverty trap, land security/tenure, food supply from local, national imported production, aid, infrastructure (roads, storage, water), food hoarding	Trade restrictions, debt repayments
Social	Population growth, poor health and reduced labour (especially HIV), war: deliberate food destruction, gender equity	War and corruption, refugees and displacement, rise of middle class and changed food tastes
Environmental	Natural disasters of drought, desertification, floods, pests, overcropping and overgrazing, urban sprawl	Pollution and climate change especially drought

Since 1992, the proportion of short- and long-term food crises that can be attributed to human causes has more than doubled, rising from 15 to over 35%.

Research activity

Fifty years ago, most counties in Africa were more than self sufficient in food and yet now most are major importers and reliant on outside food aid. By 2008, the FAO warned that 27 sub-Saharan countries needed urgent help with food security. Gather evidence for Africa's systemic problems, based under the headings of:

Economics (underinvestment, trade, aid, debt)
Politics (government corruption and war)
Health (HIV/AIDS)
Demographic (natural increase and migrations)
Environmental (hazards and resources)

Key references: Famine Early Warning System: **www.fews.net/Pages/default.aspx**
World Food Programme: **www.wfp.org**
UN Environment Programme: **www.unep-wcmc.org/habitats/drylands/index.htm**
International Food Policy Research Institute: **www.ifpri.org/**

Population pressure and resource relationships

Academics are split between two opposing views on the relationship between population and resources, as summarised in Figure 30. However, both views may be appropriate at different scales.

- On a global level the growing suffering and famine in some developing countries may support a Malthusian scenario.
- On a national scale some governments have been motivated by increasing populations to develop their resources to meet growing demands, as in Mauritius. The Green and Gene Revolutions may also be used as evidence of 'technology fixes' to solve food insecurity especially famine. This idea underpins Boserup's theory.

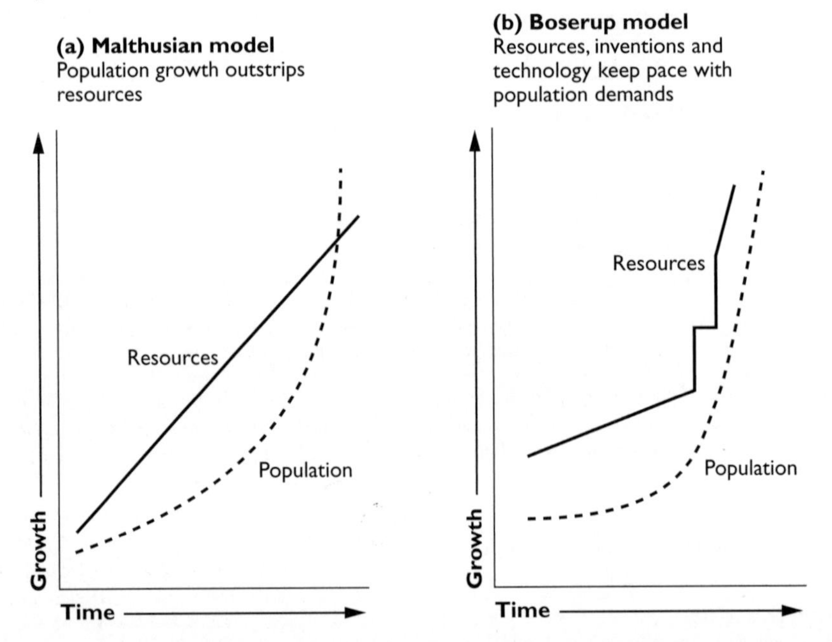

***Figure 30** Two models of the relationship between food supplies and demand*

For decades, agricultural science has focused on boosting production through the development of new technologies. It has achieved enormous yield gains as well as lower costs for large-scale farming. But this success has come at a high environmental cost. Furthermore, it has not solved the social and economic problems of the poor in developing countries, which have generally benefited the least from this boost in production.

Any farming technique, whether on land or using fresh and marine waters, will have a negative impact on natural systems. This ranges from modification to total destruction. Farming may be seen as a system, with the inputs of technology creating detrimental outputs as well as food (Table 14).

Table 14 Farming inputs and outputs

Input	Purpose	Outputs — environmental impacts
Machinery, e.g. tractors and combine harvesters, and air transport for perishables	To replace human or animal labour, and increase efficiency	Pollution: increased fossil fuel use for transport and refrigeration. Increased food miles and packaging. Soil compaction and erosion. Loss of biodiversity in wetlands, hedgerows, forests. Trawling impacts on marine ecosystems and non-target species, e.g. dolphin, albatross
Chemical fertilisers, especially nitrogen products	To increase yield, by providing high levels of plant nutrients	Pollution: eutrophication of water by agricultural runoff, loss of biodiversity, e.g. sugar cane in Caribbean
Pesticides	To remove insects and other pests, which could reduce yields	Toxic chemicals, especially DDT, entering the food chain and causing damage to organisms, which were not the intended 'victim' of the chemical
Herbicides and fungicides	To remove weeds, which take up space and use up nutrients, and to reduce fungal diseases that reduce yields	
Antibiotics	To increase resistance to disease in livestock and fish and to increase yield	Antibiotic-resistant bacteria and the danger of epidemic outbreaks; fears for human health due to consuming meat
Animal/fish feed	To increase the density of animals/fish kept in a given area	Increased demand for food crops to be used to feed livestock, and therefore increased pressure to clear areas (e.g. forests) to produce more crops especially soya. Waste products from animal slurry are highly toxic. Methane from livestock is a potent global warming gas. Intensively reared cattle are fed diets rich in protein and energy. For every acre of feedlot in the UK two more are farmed overseas to meet its needs

Any type of organic farming has relatively low impact on the environment compared to large-scale intensive or extensive commercial ventures. These have dominated modern farming in westernised countries since the Second World War, and more recently developing countries have intensified agriculture with the Green Revolution. In the EU, reforms of the Common Agricultural Policy now foster more environmentally friendly agriculture, with a growth in LEAF farms (Linking Environment with Farming). However, organic farming tends to be less profitable than more technologically based types, and has suffered because of the global recession of the early twenty-first century. It must be remembered that the majority of farmers in developing counties are still rooted in subsistence and small-scale production methods, which are often also organic because of poverty.

Enquiry question 3

What is the role of desertification in threatening life at the margins?

Scale and impact of desertification

Desertification is defined by the UN Convention to Combat Desertification (UNCCD) as 'land degradation in arid, semi-arid and dry sub-humid areas resulting from various factors, including climatic variations and human activities.' Land degradation means the reduction or loss of the biological or economic productivity of drylands. 70% of drylands are classed as degraded (excluding hyper-arid deserts) and suffering desertification.

Drylands include all terrestrial regions where the production of crops, forage, wood and other ecosystem services are limited by water. These are places where the climate is classified as dry sub-humid, semi-arid, arid or extreme hyper-arid (where few people live). The precipitation amounts are often unreliable and sporadic, which makes ecosystems fragile and vulnerable to over use. Drylands cover about 41% of the Earth's land surface and are inhabited by over 2 billion people (about one third of world population). Desertification is found on all continents except Antarctica, and affects the food security of over 2 billion people. Africa and Central Asia are particularly at risk, with three key areas of vulnerability in sub-Saharan Africa: the Sahel, the Horn of Africa and southeast Africa. Severe droughts occur on average in Africa once every 30 years, although the major cause here is human mismanagement.

Causes of desertification

The Millennium Ecosystem Assessment identifies the cause as the long-term failure to balance human demand for ecosystem services and the amount the ecosystem can supply. Pressure is increasing on dryland areas for food supply for people and animals, fuelwood, building materials and water. This increase is linked to a combination of factors shown in Figure 31.

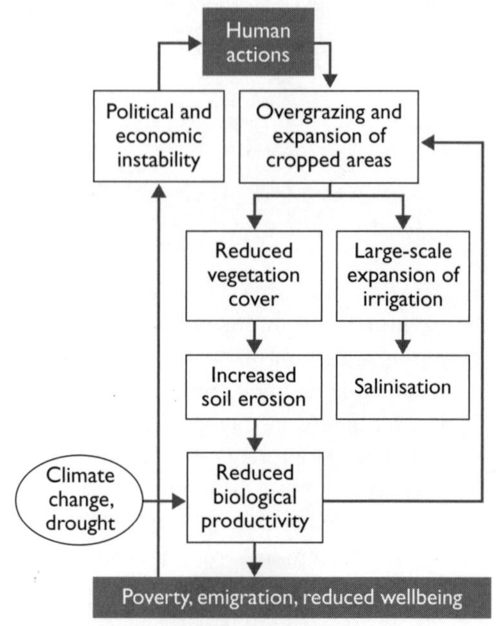

Figure 31 Simplified causes of desertification

The characteristics and vulnerability of dryland ecosystems

Drylands are characterised by a water deficit. This constrains two major interlinked ecosystem services: primary production and nutrient cycling, which directly govern both natural and agricultural systems. Crop production, forage for animals and fuelwood supplies are affected. The level of aridity partly determines whether nomadic or sedentary agriculture is practised. In more developed regions sophisticated irrigation systems may allow more intensive agriculture, but most dryland regions have extensive forms of food production.

Desertification has both direct impacts on non-dryland ecosystems and indirect impacts on people outside the actual regions, through dust storms and advancing deserts. Droughts and loss of land productivity are major factors in creating environmental refugees.

Research activity
From sources like UNEP, the UNCCD and Greenfacts, build a factfile on at least two desertified areas from contrasting economic areas on the characteristics of dryland environments. Use headings suggested in Table 11. Ensure you include the vocabulary of: hyper-arid, mineral crusts, flash floods, net primary productivity, food chains, resilience, xerophytes and pyrophytes, role of producers, consumers and decomposers, e.g. termites.

Key references: **www.unccd.int/** and **www.greenfacts.org/en/desertification/**

Food production, supply and desertification

Food production generally requires massive amounts of water, for example to grow 1 kg of wheat needs 1,000 litres of water, and to grow 1 kg of rice 3,000 litres. Irrigation can ensure an adequate and reliable supply of water, which increases yields of most crops by up to 400%. Although only 17% of global cropland is irrigated, it produces 40% of the world's food. Increasing irrigation efficiency and limiting environment damage caused by salinisation, damaged aquifers or reduced soil fertility are important for ongoing food availability.

Research activity
Find out why salinisation is a threat to food production in drylands, and the areas most affected globally. What techniques are successful in reducing this threat?

Key references: UN and Greenfacts for desertification.

The Millennium Ecosystem Assessment of drylands:
www.millenniumassessment.org/documents/document.291.aspx.pdf
African case studies: **www.drylandsresearch.org.uk/**

The sensitivity to human pressure increases with aridity. Human population in drylands increased by 18.5% between 1990 and 2000. Forty-three percent of land in Africa is classed as drylands, supporting 45% of the total population (325 million). Many African countries suffer from desertification and frequent reccurrence of drought but are also vulnerable because of their poverty. Natural resources are critical for survival, land degradation occurs due to population growth, low technological development and unsustainable resource use. Many African countries suffer from weak institutional and legal frameworks and corruption, and lack the capacity to deal with the problems of desertification and drought.

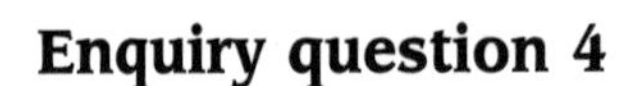

Enquiry question 4

How effective can management strategies be in sustaining life at the margins?

Strategies and techniques to increase global food supply and security

Increasing the food supply differs from security of supply. Both are difficult to achieve, but globally, increasing food supply has been given the priority. A **strategy** means the overall aims and tactics of a scheme, and it is implemented by various delivery **techniques** involving a range of policies and actual technology. Tackling insecurity means looking at all aspects of the food system from field to table and back again.

Figure 32 shows the stages required to improve food security and Figure 33 summarises the range of strategies that might be used in the quest for food security.

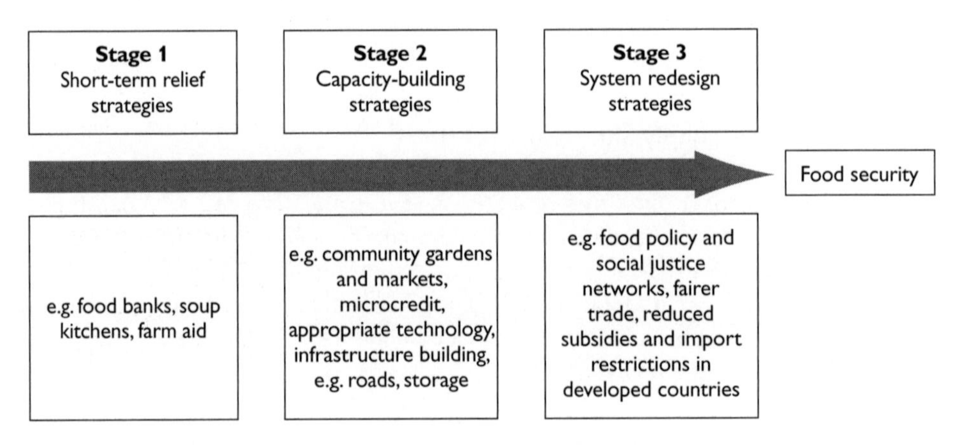

Figure 32 Stages in tackling food insecurity

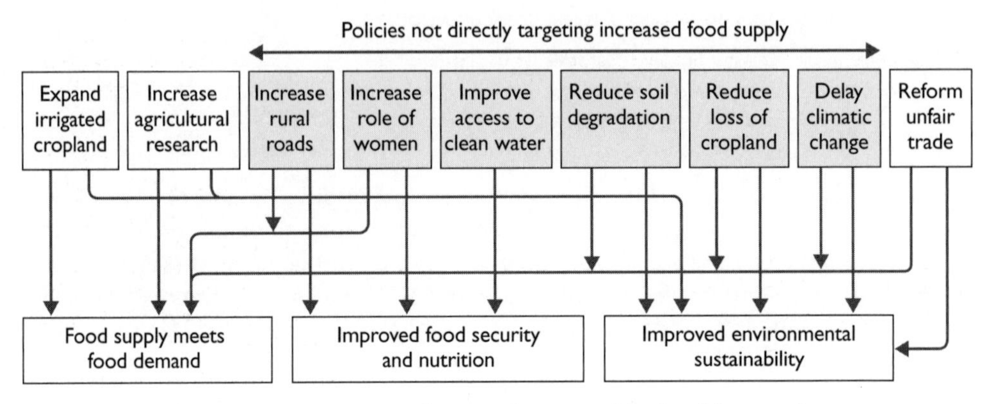

Figure 33 Spectrum of strategies to tackle food insecurity

Food security and trade liberalisation

Increasing food production leads to greater availability of food and generates income, which can help break poverty cycles. Trade liberalisation is opening up markets slowly, but there are costly barriers to overcome. The Doha round of talks by the World Trade Organization has so far failed to make trade rules fair, encourage trade liberalisation and assist developing countries to participate in the global trade environment. Until there is fairer trade on a global scale, rather than the small but growing Fair Trade movement spearheaded by NGOs, food security is unlikely.

Players and the need for more international efforts

There are a whole range of players from individuals to global institutions involved in the delivery of strategies (Table 15).

Table 15 Players in the delivery of strategies

Player	Role in sustaining life of the margins	Examples
Individuals, e.g. farmers	Direct producers of food Communities harbour stores of valuable local knowledge, coping strategies and innovation Their cooperation is critical to ensure environmental sustainability	Role of Fair Trade organisations and impact on individuals Local organic farms, LEAF
Government	Provides funding for agricultural research and development Important in creating political and economic conditions to ensure stability of food supply Response during times of crisis	Large-scale rehabilitation projects, e.g. China's Great Green Wall, Japan or UK overseas aid projects Legal and conservation frameworks
Businesses and TNCs	Research and investment into new farming methods and technologies Resource exploitation and trade in cash crops, fertilisers and farm machinery Profit motive	The development of GM crop varieties such as Golden Rice Monsanto TNC
NGOs and foundations	Community level support for farmers in the developing world Education, training and skills providers Many promote social equity, for instance female empowerment	Implementation of sustainable dryland farming in Sudan by Practical Action Emergency aid, e.g. Oxfam Toyota Motor Corporation The International Alliance Against Hunger
Research organisations	Conducting scientific research on new crop varieties and farm systems Not for profit motives Education and skills training of farmers	The development of HYVs by the International Rice Research Institute AGRA's work on a Green Revolution in Africa

Player	Role in sustaining life of the margins	Examples
IGOs such as UNEP and FAO	Promote international cooperation Implement global actions such as the Millennium Development Goals Monitoring and research to identify problems and seek solutions Development assistance and aid to the developing world	World Bank's Global Response Food Programme 1994 UN Convention on Desertification
Watchdog pressure groups	Research and information gathering Lobbying of agencies	World Resources Institute USA Coalition Food SUSTAIN

Sustainable strategies

There are three overarching aims if management is to be sustainable:

- supply of food should meet the likely increase in food demand
- food security for all
- environmental sustainability for long-term security

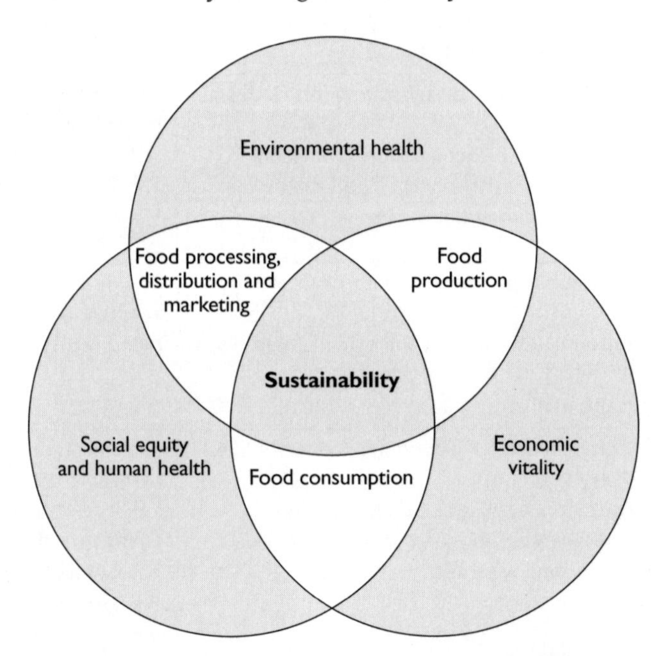

Figure 34 The basic criteria for sustainability

Figure 34 shows the overlapping nature of the food system linked to the three basic criteria needed for sustainability. A well-functioning food system not only improves human health and social wellbeing, but also, in the long term is positive for the environment and the economy and allows them to cope with shocks from natural and human created disasters.

Overcoming food supply and security requires a range of different techniques. Table 16 shows combinations of techniques in any one policy.

Table 16 Different techniques for overcoming food supply and security

Immediate relief	**Longer-term international response**	**Diversification**
Examples: World Bank Global Food Response Programme; Reliefweb food parcels	Example: UN Convention to Combat Desertification helping prepare national strategies	Example: aquaculture: freshwater fish ponds for, e.g. Tilapia in Malawi; also important for protein
Bottom up	**Top down**	**Low technology**
Favoured by NGOs, often low technology more appropriate to local area, but may not cover a wide area	Where national strategies are run by government organisations — often involving prestigious high profile, high technology projects, e.g. megadams	Example: Sahel eco-farm approach: intercropping, microdosing of fertilisers into crop planting holes
Technocentric		
From dams to e-technology. Increasingly innovative ways of using mobile phones in even remote and poorer areas: mobiles are used to find the best markets for perishable crops like bananas and fish. At an institutional level: UNICEF has had recent success with the text messaging Rapid SMS system in monitoring and delivering the protein-rich ready-to-use therapeutic food, Plumpy'nut, in drought-hit Ethiopia		

Research activity
Collect case studies from a range of organisations and scales, including more top down, often technocentric schemes like China's Green Wall afforestation, ICRISAT in India. Also bottom up, smaller scale schemes such as Sahelian eco-farms, Farm Africa, the World Fish Center. Evaluate their success in increasing food security by setting clear criteria.

Key reference: **www.droughtnet.org**

Sustainability: eat well and save the planet?

Rather than the 'business as usual' scenario, with little thought for future generations, sustainability is about seeking ways of providing food, water and energy that are long-lasting and have less of an impact on the environment. Sustainability includes:

- all aspects of the food chain from production, transportation and processing to its delivery on to a plate
- 'local' — minimising energy used in food production, transport and storage

Places like the UK are effectively at the top of the global food pyramid, and play a vital role in developing global food security, especially in a century dominated by fears of climate change.

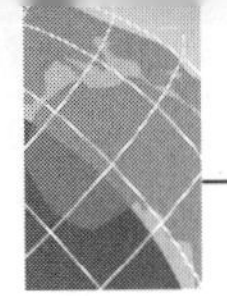

Research activity
How would the ethical and environmental policies suggested by pressure groups such as the Fairtrade Foundation and Sustain help achieve greater food security? Is there any evidence that governments are taking on such policies? You could look at the UK's Sustainable Commission and Department for International Development websites and the impacts of palm oil development for food and biofuels.

You also need to investigate the increasing threat of climate change to sustainable food supplies and security. Starting points would be Oxfam, the Famine Early Warning System and World Food Summit websites.

Key references: Food sustainability: **http://news.bbc.co.uk/2/hi/in_depth/629/629/7148876.stm** and **www.sustainablefood.com/**

NGO, e.g. Practical Action in dryland areas: **http://practicalaction.org/?id=region_southern_africa_reducing_vulnerability**

The world of cultural diversity

Introduction

One world, over 6 billion people, more than 200 nations, 6,912 living languages, many cultures and even more subcultures. One of the growth subdivisions within geography is that of **cultural geography**, the study of the characteristics of culture.

Here are some of the topics you will study in this option:

- Punk
- Chavs and Goths
- Hip-Hop
- McDonaldisation and Starbuckisation
- Role of companies like Walt Disney and Microsoft
- Mall culture
- Compulsory Welsh

Since the mid-twentieth century, culture has changed faster than ever before with the growth and flows of technology, media and globalisation. All urban and many rural landscapes are the product of culture. The exploitation, vulnerability and management of landscapes, i.e. places, are often the result of cultural decisions. However, culture is a complex concept, which has different meanings to people in different places and changes over time. At present some cultures are embracing or localising globalisation, others actively resisting its associated uniformity.

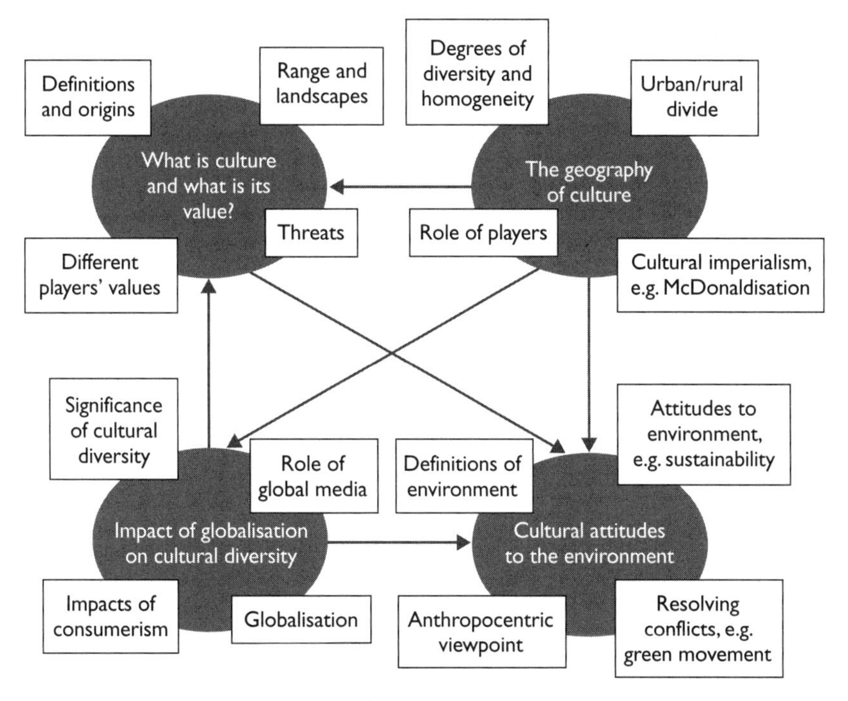

Figure 35 Topic concept map

Figure 36 The synoptic context

Research tip

While thinking of the synoptic aspects of the course, a simple checklist based on the topic/enquiry questions from the full specification will help you separate your research into manageable chunks (Table 17).

Overviews, case studies, quizzes. Key reference: **http://anthro.palomar.edu/change/default.htm**

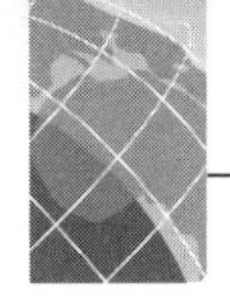

Table 17 Checklist for the world of cultural diversity

Enquiry question 1 What is the nature and value of culture in terms of peoples and places?	Enquiry question 2 How and why does culture vary spatially?	Enquiry question 3 How is globalisation impacting on culture?	Enquiry question 4 How do cultural values impact on our relationship with the environment?
• Type of culture? Sub-culture? • Type of landscape produced: ethnoscape? financescape? commodityscape? mediascape? • Pressures on the culture and cultural landscape • Is the culture/landscape vulnerable? How? Why? • How is local culture valued? Protected?	• Factors influencing degree of cultural diversity/ homogeneity • Urban? Rural? • Players involved and attitudes to cultural diversity or homogeneity • Attitude of local and national governments • Evidence of cultural imperialism?	• Significance of globalisation on area • Role of global media corporations • Evidence of globalisation? Hybrids? • Link to global consumerism	• Type of attitude to environment • Exploitation? • Preservation? • Management and mitigation? • Role of green movement • Efforts at sustainability
Key refs used	Key refs used	Key refs used	Key refs used

Enquiry question 1

What is the nature and value of culture in terms of peoples and places?

Definitions

Boris Johnson, Mayor of London, said on the BBC programme *The Culture Show*: 'Culture is what distinguishes men, human beings, from animals'. On this show, celebrities debate current cultural issues, including: *EastEnders*, sex, coffee, football and binge drinking. This type of culture may be best described as 'popular' culture rather than 'high' culture, which might be represented by art and classical music.

The very word culture suggests growth and change. It is basically a **system of shared values** in a society, which influences lifestyles and creates boundaries for behaviour and interaction with others (Figure 37). These boundaries can be seen in dress and greeting codes, religious festivals, and also in architecture and use of cityscapes and rural landscapes. Some governments, like the UK, have a separate department devoted to the development of culture.

Cultures develop norms, values, and behaviours that are suited to that environment. Over time, they take on the strength of **tradition**. Even when circumstances change, traditions often do not, since early conditioning is very difficult to overcome.

Culture is dynamic:

- It is passed on from generation to generation.

- It evolves naturally over time, particularly with increased contacts with other groups and beliefs — although individuals may perceive that their culture has not changed.
- It is open to external influences, for instance from another culture. Newcomers to an area, whether from inside a country or migrants from overseas, will either resist the culture they join or assimilate. Even third generation immigrants may retain some of their original culture. This is a two-way process where the core culture will also be changed, as shown in food tastes, religion and sometimes clothing in areas rich with migrants.

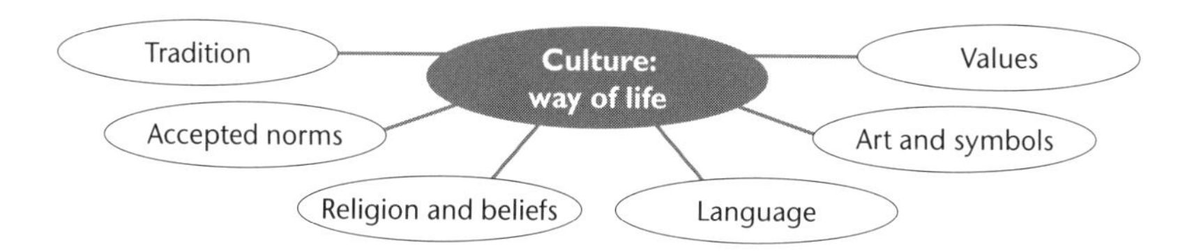

Figure 37 Components of culture

Cultures are affected by contact between societies. Contacts may be **voluntary**, as in economic migration and holidays, or **forced**, by war and refugee migration. Cultural ideas may transfer from one society to another, through:

- **Assimilation** — the adoption of a different culture by an individual or group, often quickened by intermarriage.
- **Diffusion** — the movement of cultural traits and ideas from one society or ethnic group to another. While the main form of a trait may be transferred, the original meaning may not, as in a McDonald's hamburger. These are considered cheap fast food in the USA or UK but they are more of a special occasion food in China or India, where the veggie burger has globalised the original form. Stimulus diffusion means an element of one culture leading to an invention or propagation in another.
- **Acculturation** and **transculturation** — processes by which a culture is transformed due to the adoption of cultural traits from another society. Colonisation has replaced the cultures of many indigenous people, as seen in Australia where Western culture dominated the original Aboriginal way of living. The Aborigines fiercely defended their traditional culture and are now better respected.

Groups may reject the values and beliefs of a core culture. These groups fall into two categories:

- **Separatists**, who often choose to live in their own enclaves and minimise contact with others, for example employees of a TNC in a foreign country or 'sunseeker' retirees from Britain in Spain. More settled groups include the Amish in the USA and Mennonites in Belize. Cults are specific religious groups, often living together in a community with a charismatic leader. They are often seen as potentially dangerous, unorthodox and extremist by the dominant religious organisations in a society.

- **Sub-cultures** seek to challenge or undermine the core culture and are initially seen as shocking and dangerous, but watered down versions of their 'culture' often enter the mainstream eventually. Examples are punks in the 1970s, and the eco-warriors who campaigned against road developments in the UK in the 1980s. By 2006, there were two main teenage sub-cultures in the UK: Emos, including Goths; and Chavs.

The expression of any culture is in its buildings, landscaping and attitudes to conservation and exploitation.

- Governments may foster and help maintain culture by investing in specific signposting around housing estates and tourist trails.
- Schools, community centres and even subways often have expressions of local culture in the form of painted murals, and more unofficial graffiti.
- Shopping malls have a distinct cultural expression, even if it is a cloned type of expression where a mall may look similar in India, Mexico or France.
- Different organisations, from TNCs to hospitals, schools to local authorities foster specific cultures made up of the assumptions, values, norms and tangible signs (artefacts) of organisation members and their behaviours — e.g. Japanese factories in the UK.

Research activity

Identify the aspects of culture that define the UK, and why Wales, Northern Ireland and Scotland have devolved power and separate identities from England. Look at the recent Life in the UK test for citizenship: **www.lifeintheuktest.gov.uk/** and statistics and research into British culture by the government including the 'Taking Part' survey: **www.culture.gov.uk/reference_library/research_and_statistics/default_7.aspx**

EU portal on culture in all member countries: **http://ec.europa.eu/culture/portal/activities/heritage/cultural_heritage_en.htm**

Assess how Islam is not a culture but a body of principles and universal values which can be acculturated.

Choose a minority culture suffering persecution such as the Romani and identify how such a community can retain its culture despite being scattered globally: **www.romani.org/**

Cultural landscapes

Place, as opposed to just space, is an important component of everyone's life, and is heavily influenced by cultural values. Cultural landscapes can be:

- Historic — the remains of an ancient culture's landscapes.
- Modern — a 'new' landscape reflecting the culture of today.
- Mixed — a fusion of the ancient and modern, which characterises most places today, and acts as a focal point for both the inhabitants of an area and tourists, e.g. city walls, churches and beauty spots.

Historical and beautiful landscapes are valued by most societies, which may attempt to preserve them by awarding Conservation Area status. In cities, new buildings adjacent to old ones may be designed to replicate their main forms. This can be seen in new malls that may have external architecture replicating surrounding historic features. Some countries with shorter preserved histories covet the historic sites of others.

A key player, the United Nations, designates landscapes such as the Taj Mahal, Ayers Rock or Stonehenge as World Heritage sites. More natural landscapes such as Dorset's Jurassic Coast or the USA's Wild West have the same high level designation. Even landscapes heavily influenced by people, such as the industrial and farmed landscape of the Lake District or the reforested and grazed New Forest, are fiercely defended. The latest National Park in the UK, the South Downs, has been created to protect rolling chalk hills, under threat from agricultural techniques and urban sprawl.

The process of globalisation has created more globalised forms of landscapes as well as opposition to the erosion of local identity. Cultural landscapes that cross national and cultural boundaries can be classified as shown in Table 18.

Table 18 Cultural landscapes

Ethnoscapes	Created ethnic landscapes: Curry Mile in Birmingham, Chinatown in London and San Francisco or the nine New Towns outside Shanghai modelled on different cultures, e.g. Canadian-inspired Feng Jing complete with maple trees, and Thames Town, which looks like a British village. Allotments are a current form of conservation of a local culture
Financescapes	Statuesque landscapes of corporate tower blocks and offices such as Canary Wharf in London, the 101 Building in Taipei and the Pudong financial district in Shanghai
Technoscapes	Centres of hi-tech industry, often developed with hi-tech architecture such as Putrajaya in Malaysia, Silicon Glen in California, Cambridge Science Park, UK or Grenoble in France. Technoscapes of the past may be preserved, e.g. Battersea Power Station in London and La Villette in Paris
Commodityscapes	Sites of consumption, such as the old buildings making up Knightsbridge in London, or the famous market square in Marrakech — Djemaa El Fna — now a World Heritage site. Westfield shopping mall, London, is an example of the global cloning of retail centres
Mediascapes	With clever marketing, American music, movies, television, computer games and software have 'gone global', influencing the tastes, looks, and aspirations of virtually every nation
Ecoscapes	Landscapes based on natural features, such as the Grand Canyon, USA, Antarctica or Wolong in China

Research activity

Research how different groups of people can vary in their views about such landscapes and value them very differently. See the work of a New York based anthropologist: **www.intcul.tohoku.ac.jp/~holden/MediatedSociety/Readings/2003_04/Appadurai.html**

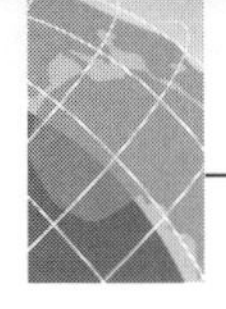

Pressures on cultures

Cultural landscapes are constantly under environmental, social and/or economic pressures to either change or be preserved, as attitudes alter and values shift in society.

Cultural **genocide** is the deliberate destruction of a group of people or nation for political, military, religious, ideological, ethnical or racial reasons, such as occurred in Rwanda in 1994. **Ethnocide** means the destruction of the culture of a people, as opposed to the people themselves, as China has done in Tibet.

On a smaller scale, but more common, are urban regeneration projects involving forced moves, seen in cities as diverse as London, Shanghai or Mumbai. The development of the Olympic Village site in Stratford and Newham was controversial and hotly debated by some local people because it may destroy the existing community and culture.

For many isolated and 'switched off' areas and countries, tourism offers a tempting route to economic development. However, it may also threaten their cultural values and landscapes, as seen in Bhutan, Macchu Picchu in Peru, and the Masai people in east Africa. An extreme form of cultural change is prostitution for the tourist trade, which has made places like Bangkok infamous. Even in the UK, tourism has created cultural changes, as seen in the rise of second home ownership in the Lake District or Wales.

Technology creates another interesting pressure: automobiles, televisions, stereos, cellular phones, computers and iPods. These have made it easier to know about the values of people all over the world.

Sectoral change in employment has led to the demise of cultures associated with rural primary production. As farming has declined in the UK, leisure and tourism have increased in rural areas. Urban influences have escalated, from counterurbanisation to the banning of fox hunting.

Research activity

Assess how culture may affect health. New diseases from new contacts, HIV/AIDS, anorexia, female mutilation, binge drinking, smoking, are all linked to cultural traits. A starting point is: **www.answers.com/topic/cultural-factors** and **www.york.ac.uk/healthsciences/equality/cultural.htm**

Protection and players

One influential global player is the United Nations Educational, Scientific and Cultural Organization (UNESCO), founded in 1945 with the rather ambitious goal to 'build peace in the minds of men'. It basically promotes international cooperation on education, science, culture and communication among the 193 countries which are member states. One of UNESCO's aims is the 'preservation and promotion of the common heritage of humanity' by World Heritage sites. This raises difficult questions about:

- which cultures and cultural landscapes should be preserved
- who should decide how and where cultures are to be preserved and promoted: a disproportionate 50% of all World Heritage sites are in Europe and North America, reflecting the influence of Western culture in UNESCO decisions

Research activity

Assess how some landscapes are seen as a commodity, some as a national symbol, and some as a threat.

Choose a country/area in which a distinctive language is being deliberately retained as part of a cultural initiative, such as Spain's Basque region, or Wales.

Assess the role UNESCO and the EU have on cultural diversity, e.g. 2008 concert celebrating cultural diversity: **www.youtube.com/watch?v=9flFbJw3pks**

EU Commission for Culture: **http://ec.europa.eu/culture/index_en.htm**

UNESCO: **www.unesco.org/culture**

Enquiry question 2

How and why does culture vary spatially?

Homogeneity and cultural diversity

The amount of cultural diversity varies greatly across the world, at both national and local levels. Some countries are more culturally homogenous than others, such as Japan, Iceland, and Saudi Arabia.

In contrast are the large-scale 'melting pots' of countries such as the USA, with large immigrant populations, and consequently a wide range of religious beliefs, customs, and values. With successive generations the original culture may be diluted. Differences still remain between rural and urban cultures: African American from European, East Asian from South Asian, and religious believers from secularist.

Research activity

Contrast and compare the following policies on cultural diversity.

UK government site: **www.homeoffice.gov.uk/**

Blog and information site for foreigners in the UK: **www.foreignersinuk.co.uk/contacts_119.html**

Japan's culture and migration policies: **www.migrationinformation.org/Profiles/display.cfm?ID=487**

And conversely, the cultural assimilation of Japanese immigrants into Brazil: **www.nowpublic.com/culture/100-years-japanese-immigration-brazil**

The TNC Nissan gives an interesting viewpoint on cultural diversity: **www.nissan-global.com/EN/COMPANY/DIVERSITY/CULTURE/index.html**

Cultural diversity in rural and urban areas

The key 'switched on' locations in the world are certain cities where success has bred success. These gateways or hubs automatically attract not only large numbers of immigrants but also the youthful and vibrant people most eager to assimilate change. Hence cultural diversity is inevitable, as seen in two key world cities: London and New York, or the megacity of Los Angeles.

Rural areas globally tend to have lower ethnic diversity because of fewer employment opportunities and greater isolation, especially those areas that are 'switched off' from our globalised world. Rural communities may be less welcoming to different cultures, and whether rich or poor, immigrants are often classed as outsiders. Grants are available in many countries, such as the UK, to improve cultural facilities especially based on the arts, e.g. the Rural Development Council.

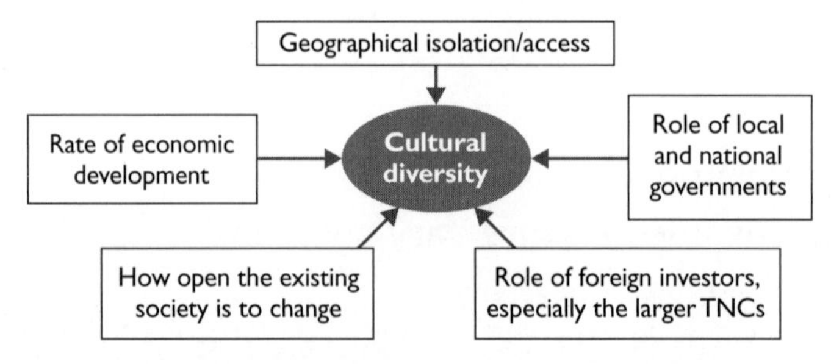

Figure 38 Factors affecting cultural diversity

Research activity

Investigate the causes and consequences, including the spatial distribution, of cultural diversity in a world city such as London compared with smaller cities and rural areas.

Key references: multiculturalism — Joseph Rowntree Research Foundation: **www.jrf.org.uk/knowledge/findings/housing/1950.asp**

Interculturalism using London and Bristol as specific case studies: **www.guardian.co.uk/graphic/0,5812,1395103,00.html**

www.nyidanmark.dk/bibliotek/publikationer/rapporter/uk/cultural_diversity/index.htm

Attitudes towards cultural diversity

Not all cultural and political groups agree that cultural diversity is positive. Tolerance of others is not universal and the host country may reach a tipping point when the carrying capacity, either real or imagined, is reached. Recent changes to immigration policy in the UK suggest this threshold has been reached. Attempts to help assimilation of other cultures into mainstream culture may be seen in education policies.

Diversity may also be seen as a threat to local culture, as happened in Spain when Franco was in power and banned the use of the Catalan, Basque and Galago languages. Regional languages are seeing a revival with promotion of diversity, e.g. Breton, Welsh, Gaelic and Catalan.

Research activity

Contrast the policies of the UK and France on immigration assimilation.

Tibet and Israel provide contrasting case studies of viewpoints on cultural landscapes:

www.international.ucla.edu/article.asp?parentid=2732

http://mondediplo.com/2008/05/09tibet

Cultural imperialism

This is the promotion or imposition of a more dominant culture on to a smaller or less important one. This has happened throughout history after war and conquests of new territories — most recently in the twentieth century after the dismantling of the old colonial system and a rise in Western, especially European and North American, power.

McDonalisation is the term given to the extreme form of consumerist culture originating from the USA. Western culture generally has:

- an emphasis on high levels of resource consumption
- a high reliance on globalisation
- a political system based on 'democracy', although in differing forms
- a 'technological fix': a reliance on, and belief in, technology
- a high degree of influence from global media corporations (BBC, CNN), television from the USA, Hollywood films and global brands like Coca-Cola and McDonald's.

Supporters of the Western cultural model would describe this as modernisation or development, and a belief that this culture represents progress, rather than cultural imperialism.

Counter movements from other cultures, especially green movements, and some religious groups would disagree, as shown in demonstrations at every World Trade Organization or G8 and G20 meetings.

Research activity

Research the mechanisms of the process that has become stereotyped as 'McDonaldisation' and the counter cultural movements that it has given rise to:

www.globalisationguide.org/index.htm

www.thehumanist.org/humanist/articles/essay3mayjune04.pdf

www.wsu.edu/~amerstu/pop/cultimp.html

Enquiry question 3

How is globalisation impacting on culture?

Viewpoints on globalisation and cultural diversity

There is a debate between those who feel that globalisation will ultimately destroy cultural diversity and those who are sceptical about its power to do so (Figure 39).

Reduction of cultural diversity with globalisation	Process of globalisation forces countries and governments to adapt and change in uncomfortable ways	Globalisation is not new or global: maintains switched on areas of Europe, North America, Japan and excludes switched off places — most of sub-Saharan Africa. The rich get richer
Pessimistic hyperglobalisers	**Transformationalists**	**Sceptics**
Increasing power of TNCs. The rise of a global consumer culture and loss of local and national identity	Formation of the EU super bloc to maintain Europe's power. USA and UK reaction to the 2008 global financial crisis by nationalisation, bailouts, subsidies — a reversal of decades of a market led economy	85% of world trade is still between developed countries. Increasing divide between rich and poor

Figure 39 The globalisation debate

The role of global media corporations

Media corporations have become major players in spreading certain cultural values, and many would argue that they have disproportionate power. The companies involved, such as Disney, Viacom, Microsoft, Vivendi, argue that no one is forced to buy, watch, listen to or use their products. However, they do monopolise the view of the world for billions of people, and a matter of concern is that recent mergers mean less competition and a uniformity of coverage even to the point of reducing the diversity of languages.

Globalisation and hybrid localised culture

The idea that globalisation will produce one global culture is too simplistic. New cultural forms and landscapes as well as new languages are emerging as economic migrants increase and as the internet and texting become common even in remote areas. There are now distinctive hybrid globalised forms of fashion, music and film.

Modern technology can be seen not only as a threat to traditional forms of culture but also as a cultural opportunity. Websites such as YouTube have encouraged the global village in a virtual medium. Free social messaging websites like MySpace and Facebook, together with blogs, and the availability online of television, news and film, mean that it is possible to keep in touch with family and friends and expand cultural contacts far more easily and over a wider spatial extent than ever before. This may work to preserve at least some aspects of cultural diversity.

Research activity

Identify the role sites like Xiaonei, the equivalent of Facebook in China, have on cultural exchange. Find the reasons why Google/YouTube are banned in countries like Turkey, Thailand, Iran and China, and how this may be linked with fears about erosion of local cultures. Investigate the globalisation of cultures by looking at why Bollywood plays a large role in the film industry, or how popular game and reality shows have been marketed worldwide but subtly changed for local demands, e.g. *Who Wants to be a Millionaire* or *Pop Idol*.

The impacts of global consumerist culture

Increasing wealth and globalisation may increase choices, from basic food types to leisure pursuits. Indeed, many poorer people may choose the outward signs of globalisation for prestige and a feeling of belonging to mainstream society.

However, consumerist culture may not be best for mental or physical health. Too much choice, whether in information or brands of trainer, mobile phone or products like milk, may lead to more stress than satisfaction. Research in *Scientific American* in 2004 indicated that in affluent societies like the USA, GDP has more than doubled in the past 30 years yet the percentage of people describing themselves as 'very happy' has declined by about 5%.

Research activity

Evaluate the positive and negative effects of growing consumerism on societies in the two most rapidly developing economies: China and India.

Assess why some countries have resisted some aspects of USA culture, e.g. Starbucks in France.

Investigate the role of culture in the twinning of urban areas, such as Sichuan (twinned with Leicestershire) and Chongqing (twinned with Leicester).

Enquiry question 4

How do cultural values impact on our relationship with the environment?

Cultural attitudes and the environment

The United Nations defines the environment as, 'the totality of all the external conditions affecting the life, development and survival of an organism' which could mean a tree, polar bear or human.

Environment can be divided up into urban, semi-urban, rural and wilderness areas.

Different groups have different perceptions and place differing degrees of importance on the environment. With the present growing threat of climate change, people view even the atmospheric environment differently depending upon the country, organisation or individual. Many view environmental issues as a global, governmental problem rather than an individual, local one, hence the emphasis of the Rio Earth Summit of 1992 on its Local Agenda 21 of 'think global, act local'. Cultural views can vary from sensitive to indifferent.

Several factors affect current cultural attitudes to the environment:

- rising net income
- increased education
- personal involvement/impacts from pollution, soil erosion, etc.
- rise of local branches of international green groups such as Greenpeace, WWF
- role of media: television for information, and mobile phones and the internet to mobilise behaviour

- role of government in disseminating information and allocating resources to environmental protection: from pollution control to National Park designation. The international image of a country also now depends on its attitude and effectiveness in environmental quality, especially as foreign direct investment and TNCs can be very selective in such a globalised world
- role of IGOs, e.g. UNESCO and UNEP in World Heritage Site and Biosphere Reserves

By the start of the twenty-first century there were four main cultural views of the environment:

- **Technocentric view** — the historic Western idea that the environment is a source of resources for humans to use and a sink for waste.
- **Accommodation view** — the view of most Westernised countries at present. Improve the existing system through finance, technology, smart techniques — no fundamental change.
- **Social ecology view** — a new system is needed with community and local focus, ethically maintained, e.g. recycling, organic farming, solar energy. Ecotowns lead the way.
- **Deep ecology view** — going back to pre-industrial values. Nature dominates; careful human use of natural material; humans are part of the environment.

These viewpoints will probably all be represented in large democratic countries and may well change over time. People and societies constantly change in their cultural outlook.

Many governments, even originally exploitative ones like the UK and Australia, are moving closer to environmental cultural views that enhance ecological awareness, urban lifestyle quality and fairer societies.

China, the world's most populous state is a huge consumer of resources and pollution generator. At present, environmental attitudes in China are shaped first by tradition and culture, and second by new conceptions of economic growth and consumption. There is a cultural history of humans trying to control nature, and exploit it. But there are signs of a backlash against the immense environmental degradation resulting from China's rapid economic transformation into an industrial superpower.

Research activity

Investigate the recent changes in attitude to the environment in China with the Beijing Olympics 2008 as a possible catalyst: **http://money.cnn.com/magazines/fortune/fortune_archive/2007/05/14/100024847/index.htm**

Summarise the main findings of the EU Eurobarometer survey of 27,000 people in 2007 about attitudes and knowledge of the environment, comparing it with data from 2004: **http://exploreourpla.net/global-warming/reports/eurobarometer-attitudes-of-european-citizens-towards-the-environment.html**

Valuing landscapes

A major part of the environment is landscape, which may be valued for a series of reasons. As mass tourism and eco-visits increase, protected landscapes are increasingly seen as a marketable attraction with spin off multiplier effects in the economy.

People's cultural perceptions of landscape value determine how land and resources are used or abused:

- landscape as sacred and preserved
- landscape as pleasure — conserved, but may become degraded and polluted
- landscape as life and protected
- landscapes have resources to use for profit and can be degraded and even destroyed

Protecting areas is no longer the preserve of more affluent societies that can afford to choose not to exploit an area and protect it as a nature reserve or National Park. The first National Park was in the USA — Yellowstone, 1872. Worldwide there are now National Parks in almost 100 countries and China has 187 — all designated since 1982. Most include some degree of use as a leisure hot spot and provide a 'natural' cultural outlet. See: **http://earthtrends.wri.org/maps_spatial/maps_detail_static.php?map_select=482&theme=7**

Figure 40 shows a model based on the Kuznet environmental curve of how attitudes vary over time to the protection of the environment.

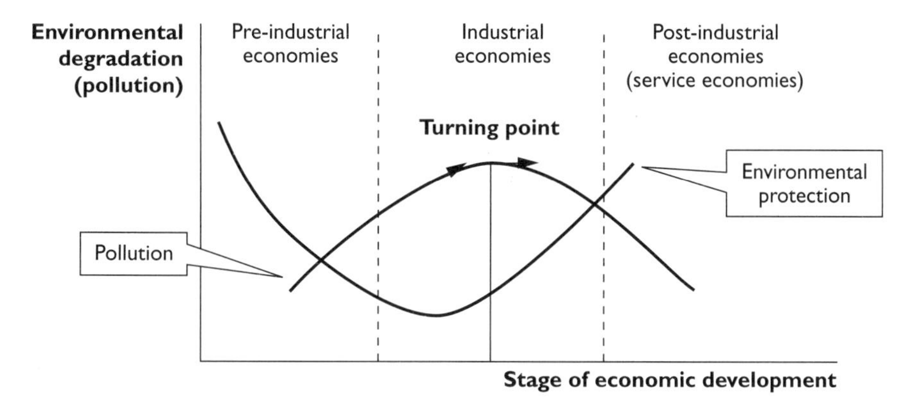

Figure 40 Attitudes to environmental protection

Sustainable environments

Sustainability is difficult to define, and is used by different cultures in different ways. The concept began with late twentieth century concerns over misuse of the Earth's resources and the increasing development gap. Sustainable development essentially seeks to generate economic progress that is equitable, will last, involves ordinary people in decision making and does not needlessly damage the environment.

According to the UK Sustainable Communities Plan: 'sustainable communities are places where people want to live and work — now and in the future. Culture can play a role in building sustainable communities.'

However, sustainability is often used as more of a 'green' concept focusing on environmental issues rather than social equity and public participation. Globally, local authorities and businesses stress 'environmental sustainability'.

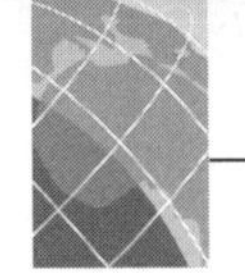

The biggest global organisation setting standards to encourage universal agreements on ethical issues is UNESCO. It has set conventions to address the fact the 'world urgently requires global visions of sustainable development based upon observance of human rights, mutual respect and the alleviation of poverty'. The EU is a signatory, as is China.

Research activity
Look at the UK government's official policy on sustainability and contrast it with that of a different cultural tradition, e.g. China or Morocco: **www.defra.gov.uk/sustainable/government/what/index.htm.**

Why do big TNCs like Shell or companies like M&S market themselves as 'green'? Can they ever be really green?

Research UNESCO's viewpoint on the role of culture in sustainability.

Consumer culture

Parents have an in-built desire to ensure a better lifestyle for their children. In Western culture this has developed into a belief that greater material wealth means more secure and happier people. Counter-cultures such as the eco and environmental movements dispute not only this but also the anthropocentric values underpinning consumerism — that humans should dominate the planet because they are more important than anything else.

Environmentalism versus consumer capitalism

Most people now live in a world dominated by capitalism, and paradoxically the same people who are profligate consumers are concerned about environmental issues — global warming, increased hydro-meteorological hazards like desertification, and biodiversity loss.

Capitalism is an economic and political system that almost inevitably leads to destruction of the environment because:

- Short-term timescales for profit put environmental protection in second place.
- Businesses are accountable to their shareholders, rather than to the wider public.
- It marginalises poorer people, increasing poverty and desperation, which leads to short-term destructive use of the environment.

There are few signs that consumption is becoming less popular, indeed it has spread to the growing superpowers of China, India and the Gulf States.

Research activity
Investigate the growing demands of countries like China and poorer developing countries on global resources.

Examine the conflicts between environmentalism and consumption by finding out how contraction and convergence could help reduce ecological footprints.

Pollution and human health at risk

Introduction

The largest global organisation devoted to health risk is the UN World Health Organization (WHO), which stresses: 'Better health is central to human happiness and wellbeing. It also makes an important contribution to economic progress, as healthy populations live longer, are more productive, and save more'.

Health is 'a state of complete physical, mental, and social wellbeing and not merely the absence of disease or infirmity' (WHO). There are many complex interrelated factors influencing **health risk** (Figure 41), which include the threats that cause both disease and disability or **morbidity**, and **mortality** (death).

Critical is the level of socioeconomic development and personal status, which will influence environmental conditions and how the risks are managed. **Global health risk** means health problems that are so large they have a global political and economic impact or **burden**.

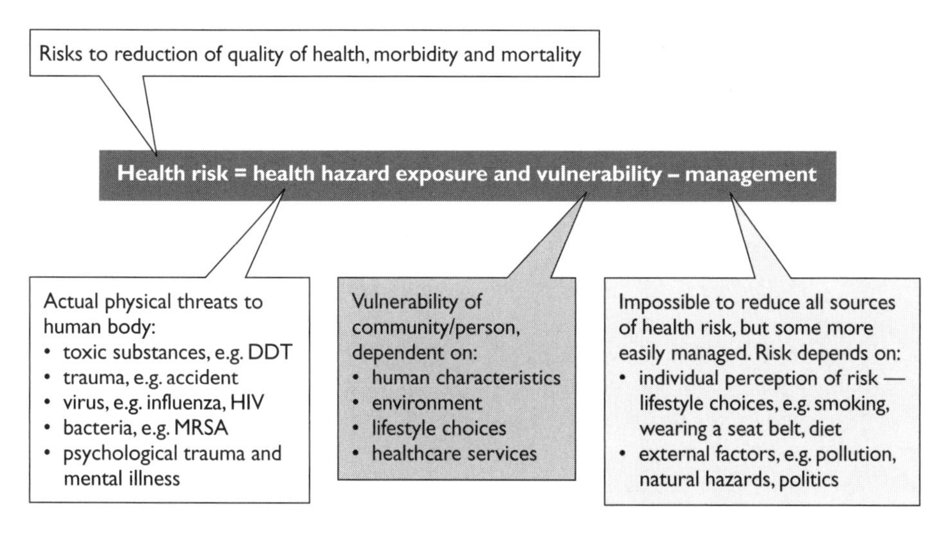

Figure 41 The health risk equation (shaded boxes highlight the geography of this topic)

You will study the patterns and trends in health risks over time and space, both globally and locally, and evaluate the factors involved, including globalisation, global shift of environmental pollution, management and players. Some of the work you have done at AS on Going Global will help here.

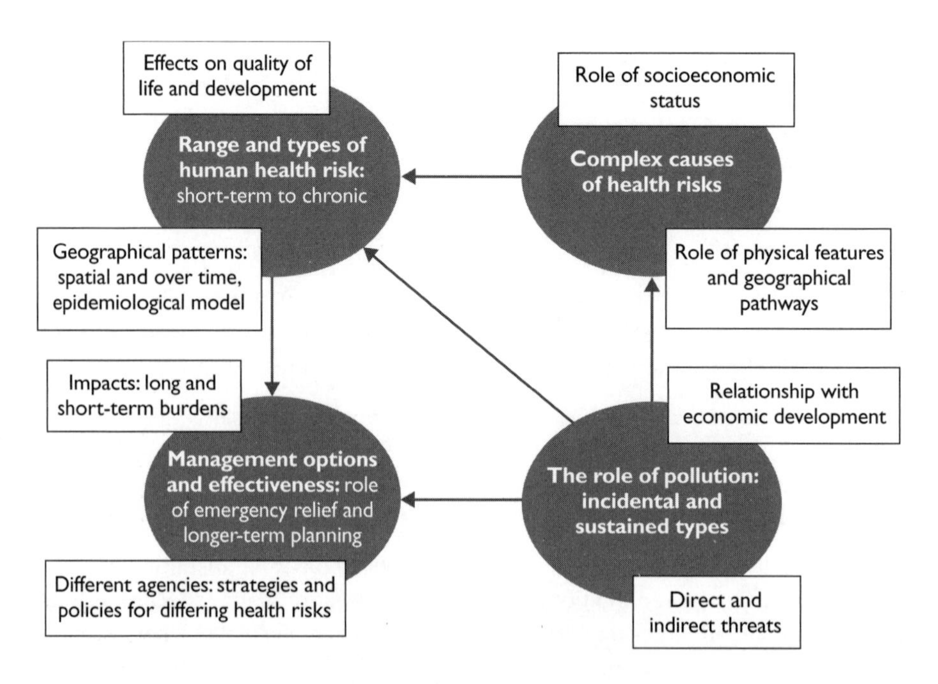

Figure 42 Topic concept map

Research tip
When you research case studies and smaller examples keep notes on all the enquiry questions, with a log of sources that you can review once the research focus pre-release material is available. You will be expected to quote some key sources in your actual exam: from books, publications, newspapers, websites, DVDs, podcasts, pamphlets etc.

Create your own graphs and comparisons between countries. Key reference: WHO, with factsheets, databases: **www.who.int/infobase/comparestart.aspx**

One key point to remember when researching this option is that you must apply a geographical focus. The specification suggests that a synoptic view of the people and power involved in any health risk must always be linked to place (Figure 43). Do not get sidetracked by medical details unless they directly link to geography.

Figure 43 The synoptic context

Research tip
While thinking of the synoptic aspects of the course, a simple checklist based on the topic/enquiry questions from the full specification will help you separate your research into manageable chunks (Table 19).

Table 19 Checklist for pollution and human health at risk

Enquiry question 1 What are the health risks?	Enquiry question 2 What are the causes of health risks?	Enquiry question 3 What is the link between health risk and pollution?	Enquiry question 4 How can the impacts of health risk be managed?
• Short-term or longer-term risk? • Infectious/ transmissible or chronic/ degenerative disease? • Pandemic, epidemic or endemic health risk? • Geographical pattern at global/ national/local scales? • Temporal pattern • Where on the epidemiology model and WHO transition model? • Social, economic, environmental, long- and short-term impacts	• Range of causes • Long-established (older) risk or emergent or re-emerging risk? • Relationship to socioeconomic status • Is it classed a health shock? • Links to any geographical feature/pathway • Influence of environmental change • Model, e.g. diffusion	• Type of pollution? • Source? • Direct and indirect effects? • Incident or sustained pollution • Link with economic development • Where on models — environmental. Kuznet curve, epidemiological model/WHO, transition model? • Evidence of pollution fatigue in reducing health risk	• Socioeconomic and environmental impacts: short- and long-term burdens • Management strategy/policy — why used and how effective? • Agencies involved • International efforts and aims • Role of long-term sustainable or short-term emergency relief
Key refs used	Key refs used	Key refs used	Key refs used

Note: enquiry question 3 will only apply if the health risk is linked to pollution, unlike risks such as diabetes and HIV/AIDS.

Enquiry question 1
What are the health risks?

An incredibly diverse range of health risks reduces quality of life and longevity. The key to success in this option is to choose a range of health risks across different areas of the world to identify the geographical features and factors involved.

Research tip
Table 20 will help you choose a range. N.B. none is compulsory. You need coverage across areas/countries that have different experiences and impacts of the same risk.

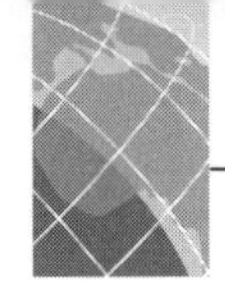

Table 20 Simple health risk classification (excluding genetic reasons)

<table>
<tr>
<td colspan="2">Infections, often communicable, often acute health risks, i.e. rapid onset or intense symptoms, split into:
• endemic, meaning a disease that has persisted for a long time in an area
• epidemic, when a disease fluctuates over time
• pandemic, when epidemics affect wide geographical areas</td>
<td rowspan="2">Degenerative, chronic health risks often resulting from longevity, not communicable:
• chronic (lasting over 3 months)
• cardiovascular (heart) diseases
• respiratory diseases, chronic obstructive pulmonary disease (COPD)
• obesity-related
• diabetes, cancer, depression
• maternal/perinatal illnesses
• degenerative
• arthritis, Alzheimer's</td>
</tr>
<tr>
<td>Vectored, i.e. involves a carrier:
• malaria or dengue from mosquito
• Lyme disease from ticks
• trypanosomiasis from tsetse fly
• plague from rats</td>
<td>Non-vectored, i.e. through bodily fluids, food, airborne, waterborne:
• HIV
• tuberculosis (TB)
• norovirus
• typhoid
• avian influenza</td>
</tr>
<tr>
<td colspan="3">Pollution created/related health risks: cholera, radiation, asthma, respiratory infections, melanoma</td>
</tr>
<tr>
<td colspan="3">Traumas: e.g. from work related accidents or transport accidents</td>
</tr>
</table>

Causes of mortality and spatial patterns of health risk

Globally there are huge health inequalities, nationally and locally, both in terms of health during life and age of death (longevity). One common measure of the level of people's health, created by WHO is DALYs (disability-adjusted life years) — the years of life spent with reduced functions resulting from health conditions of varying severity.

Table 21 Globally — what people die of based on national wealth

High-income countries (US$11,456 person^{-1} yr^{-1})	**Middle-income countries**	**Low-income countries (US$935 person^{-1} yr^{-1})**
• Low mortality countries • 66%+ of population live over 70 years • Chronic diseases dominate: cardiovascular disease, chronic obstructive lung disease, cancers, diabetes or dementia. Lung infection is the only leading infectious cause of death • e.g. UK, USA	• 50% live to 70 years • Chronic diseases plus HIV/AIDS, complications of pregnancy and childbirth and road traffic accidents dominate • e.g. China, Brazil	• High mortality countries • 25% reach 70 years • 33% of all deaths are of under 14 years. Cardiovascular diseases are the main single cause of death but infectious diseases combined (especially HIV/AIDS, lung infections, tuberculosis, diarrhoeal diseases and malaria) cause more deaths. Complications of pregnancy and childbirth also important in mortality • e.g. Kenya, Bangladesh

Globally, one in three deaths is from infectious or communicable diseases such as HIV, concentrated in poorer regions often linked with malnutrition. However, the biggest threat in developed countries is from chronic, non-communicable diseases, especially cardiovascular disease. The speed at which many countries have changed in their socioeconomic structure over the past few decades means that health managers may have to deal with both types of health risk in the same country. WHO divides the world into high and low mortality regions, correlating strongly with industrialisation and GNP/GNI (Table 21).

Research tip
Use the economic development spectrum studied in AS Going Global:
MEDCs–NICs– LEDCs–LDCs
to analyse spatial variations. Starting points are WHO's annual global burden reports. For interesting graphs and podcasts see the Gapminder website.

Health and wealth

There is an overall positive correlation between global income and health, with some anomalies like Cuba (low GNI but good health). Fifteen per cent of the world's population live in high-income countries, mainly North America and Europe, but these only account for 7% of all deaths annually. However, the simple North–South divide of the 1980s has shifted with the rapid rise of transition economies like China and the countries of Eastern Europe. By 2001 most people lived in middle-income countries, with the biggest gap ever seen between rich and poor and associated life expectancy and health risk. Since the 1990s, sub-Saharan Africa has regressed, mainly because of HIV/AIDS. Health inequalities show an inter-generational cycle, with differences beginning before birth. The scale of the problem is immense: with acute risks for the 1 billion people who live in urban slums.

Apart from national variations, there are also great regional and local variations: most cities show differences in wealth and economic welfare. In Bristol, average life expectancy may vary by 10 years between affluent and deprived neighbourhoods. Imagine the differences between an affluent gated community in Mumbai or Rio de Janeiro and a newer shanty town.

Health inequalities depend on:

- demography: age, gender, age of child bearing
- environment: housing and pollution, especially water quality
- lifestyle and habit choices: opportunities, diet, education, hygiene, job choice, smoking, drugs, drinking, exercise
- healthcare services and management: medical and infrastructure technology, access and effectiveness of regulation

Health risk patterns over time

Temporal patterns of health and disease (how health risks alter over time) are described by models like the epidemiology model (Table 22).

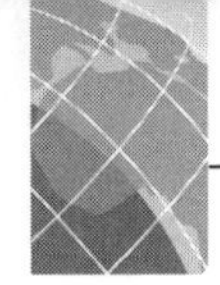

Table 22 Epidemiological model based on Omram 1971

	Stage or age			
	1 The age of pestilence and famine	**2 The age of receding pandemics**	**3 The age of chronic diseases**	**4 The age of emerging/ re-emerging infectious diseases**
Causes of health risk	Large number of infectious, acute diseases	Reduction in the prevalence of infectious diseases and fall in mortality	Degenerative and human induced diseases of affluence suffered by ageing populations	Emergence of new infectious diseases, e.g. HIV/AIDS or the re-emergence of 'old' diseases, e.g. TB, measles
Examples of types of health risk	Mainly respiratory and infectious diseases: measles, smallpox, malaria, typhoid, cholera, tuberculosis, enteritis, diarrhoea, pneumonia		Cancers Respiratory diseases including asthma	HIV/AIDS SARS Avian influenza
Link to pollution	Localised pollution, especially waterborne	Rise in all types of pollution as industrialisation increases	Environmentally conscious but consumerist society, which has pressurised governments into reduction in water and land pollution, but rise in air pollution	In low to middle income countries high rates of all types of environmental pollution
Link to economic development	Low-income countries, e.g. UK in 17th century. Currently Ethiopia, Bangladesh, although most moving to second stage	Industrialisation, e.g. UK in 19th century. Currently low to middle income countries, e.g. India, China, especially in rural western provinces	Post-industrialisation, e.g. UK in 20th century. Currently upper income countries, e.g. UK, USA, plus recently industrialising countries like urban eastern China. Ageing populations in urbanised societies	Low to middle income countries less able to cope with the double burden of health risk, i.e. infectious and chronic diseases. Early 21st century: huge rise in HIV/AIDs, smoking, hypertension, toxic effects of widespread environmental pollution not under control

The length and characteristics of the model's stages vary with local and national social, cultural, economic, and health resource factors. The fundamental shift from communicable to non-communicable diseases is predicted to be worldwide by 2020 (WHO). It began in the 1960s in low and middle income countries, with the fastest changes in 'switched on' places like South Korea, which used the experiences and more advanced technology of older industrialised nations. Stages 1 and 2 have often become compressed.

The fourth stage was added because of several unexpected events during the late twentieth century:

- The emergence of HIV/AIDS in the 1980s has seriously hindered the shift, producing a double burden of disease affecting rich (USA) as well as poor (African) societies.
- In more advanced economies, major advances in the prevention and management of heart disease and many cancers, and remarkable increases in life expectancy for the over-50s, may be called a 'post industrialisation' phase — particularly in Westernised countries.
- In many of the former Soviet republics the demise of communism led to a collapse in public health systems, an increase in tobacco and alcohol abuse and a re-emergence of infectious diseases like diphtheria and tuberculosis — supported by increasing inequity and poverty.

Research tip
Models are a useful way of summarising complex reality and helpful as a basis for comparison. You should establish where your case study is located on the model and how it may change in the future. Refresh your knowledge on how sanitation, hygiene, medical advances, etc. influence death rates as shown in the demographic transition model.

Projections are vital for health system managers and players involved in resource allocation and overall strategies. WHO forecasts that by 2030:

- Chronic disease will be dominant, even in low income countries like Bangladesh, with decreases in infectious diseases like diarrhoea, HIV/AIDS, tuberculosis and malaria.
- There will be an increase in mental illness, especially depression, which causes disability rather than death.
- Vehicular accidents may rise to third in importance, followed by obesity and smoking.

Health risk and effects on quality of life and economic development

Health risk causes a reduction of 'human capital':

- direct effect on the individual: poor health, shorter life expectancy, negative effects on children in family
- indirect effect on local/national economy: loss of work time and productivity and hence lowered GNI

Research tip
This section is effectively a foundation for the rest, once you are secure in knowing the background, the key terms and the key players then the rest will fall into place. When justifying your focus in your final report essay, show the huge range of health risk and, as a geographer, why you are concentrating on certain types and aspects.

Enquiry question 2

What are the causes of health risks?

Causes and socioeconomic status

The causes of more established health risks, such as cancer; emergent health risks such as HIV/AIDS; and re-emerging health risks, such as tuberculosis, are complex because they include:

- internal causes, including genetics, lifestyle choices by individuals, employment
- external causes, e.g. local environmental conditions such as polluted rivers, governmental strategies, role of TNCs

Figure 44 is a model commonly used by health authorities to show the direct factors (inner rings) and more indirect factors (outer rings) affecting health.

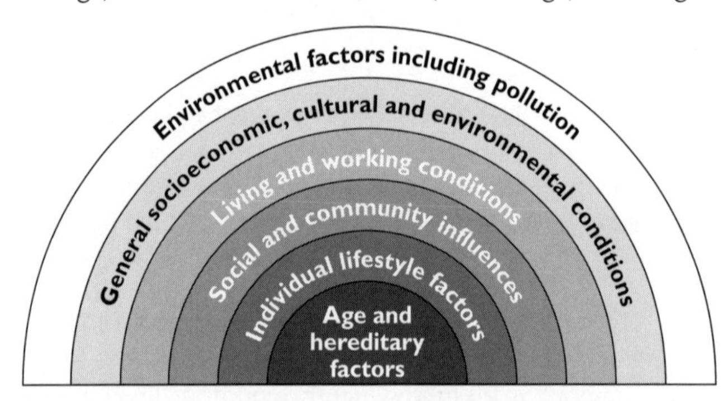

Figure 44 Direct and indirect factors affecting health

Socioeconomic status is how an individual's economic ranking relates to others in their society. It directly affects morbidity and mortality, although the relationship may depend on factors such as gender, culture and ethnicity as well as management. It varies at global, regional and local scales.

The factors behind the initial development and subsequent spread of disease may be split into socioeconomic and environmental categories (Table 23).

Table 23 Factors in the development and spread of disease

Socioeconomic factors	Environmental factors
Wealth, which often controls access to: • nutrition • better sanitation and water standards • medical and general infrastructure technology • low risk employment • higher levels of healthcare, especially immunisation • better education, and improved status for women — 'social vaccine' Migration and exposure to health risks Overuse of antibiotics and agricultural chemicals Inadequate or deteriorating public health infrastructures	Environmental changes caused by human activity, e.g. dam and road building, deforestation, irrigation, and, at the global level, climate change Natural hazard shocks, e.g. extreme heat waves and cold snaps, earthquakes and tsunami Pollution

Research activity
Health authorities within the UK have readily available data online, possibly mapped using GIS. Research the health status of your local area. See Association of Public Health Observatories: **www.apho.org.uk/default.aspx**

Geographical features and pathways

Geographical location is a major factor in the development, spread and consequences of health risk (Table 24). Infectious diseases are spread by pathogens (agents that cause disease) like viruses (e.g. measles and influenza), bacteria (e.g. meningitis and tuberculosis) and other microorganisms like parasites (tapeworms and guardia). These invade the body or contaminate the environment.

Table 24 Geographical pathways and features

Physical environment	Human environment
• Climate, topography, hydrology, geology, ecology. These may act as a barrier or conduit, or create a 'disease reservoir' for pathogens and vectors. They can be physical, e.g. lakes, or within humans themselves • Humid tropical and equatorial regions produce optimum conditions for many infectious diseases. Unfortunately, often correlated with poorest, most vulnerable countries • Changes in environment: global warming impacts	• Infrastructure: transport routes, especially air travel • Services: sewage and fresh water • Pollution pathways • Building and environmental quality including green spaces • Race, gender

Research tip
Malaria and schistosomiasis involving vectors are particularly good for showing the role of physical conditions and pathways. Meningitis, tuberculosis and measles show the role of human environments, and HIV/AIDS and SARS show communications/transport influences.

Key reference: WHO website: **www.who.int/mediacentre/factsheets/en/**

Emergent and re-emergent infectious diseases

There are repeated cycles in the rise and fall of infections. Currently there is a high rate of re-emergent diseases (tuberculosis and measles), and emergent diseases (Ebola fever, Legionnaire's and Lyme diseases, toxic shock syndrome, AIDS, hepatitis C, vCJD and a new variant of cholera).

There has also been an increase in the virulence and drug resistance of infections such as diphtheria, influenza, malaria and tuberculosis.

The links between some diseases and geographical features and pathways

The spread of many diseases, like measles, may follow a diffusion model (see Figure 45).

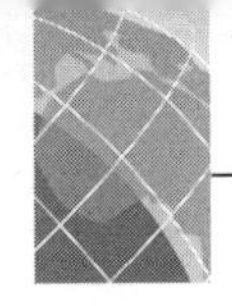

(a) **Expansion diffusion: the disease spreads from one place but remains concentrated there**

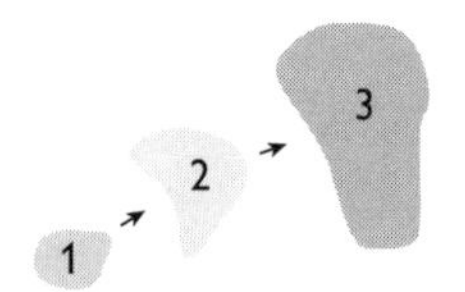

(b) **Relocation diffusion, e.g. some types of flu: the disease moves from place to place**

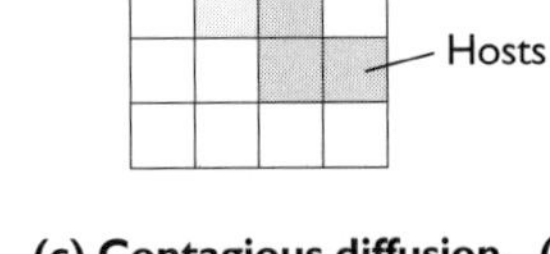

(c) **Contagious diffusion, e.g. measles: direct contact is needed to pass the disease on**

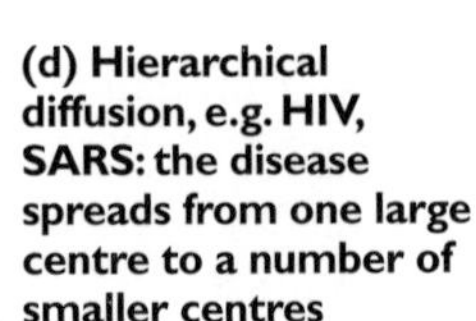

(d) **Hierarchical diffusion, e.g. HIV, SARS: the disease spreads from one large centre to a number of smaller centres**

Figure 45 Types of spatial diffusion

Research tip

Models are used by managers to control health risk patterns and GIS is an important tool. Other models you could investigate are by Wilkinson (life expectancy/income) and Bartlett who used measles to show how a disease can be endemic (remain in a population as a 'reservoir' of infection — UK, 2008).

Key reference: Science Museum — Making the Modern World: **www.makingthemodernworld.org.uk/learning_modules/geography/05.TU.01/**

Enquiry question 3

What is the link between health risk and pollution?

Environmental health is a large subset of the health risk problem. There is a range of risks associated with one-off pollution incidents, and longer-term or sustained pollution.

One-off pollution incidents may be moved by air or water along a pathway. They range from low-risk oil spills to higher-risk processed or radioactive substances, e.g. Chernobyl 1986.

Examples of longer-term or sustained pollution include:

- 1980s agricultural nitrates in groundwater linked with 'blue baby syndrome'; in 2009 the EU banned 22 commonly used chemicals in agriculture
- 1970s onwards, melanoma from CFC destruction of protective ozone layer
- climate change — fears of spread of disease
- smoking environments creating carcogenic risks — WHO aims for a worldwide ban in public places

Pollution from land, water or air sources may pose health risks. **Transboundary** effects add complexity, especially from atmospheric pollution hazards like ozone depletion and enhanced global warming. Pollution has escalated globally over the past century and increased in spatial extent with specific **hot spots** or **clusters** often created by point source incidents like Chernobyl, Bhopal and Harbin. Transition economies such as the BRIC group now suffer the most risk. Many older economies have reduced their land and water pollution rates dramatically with deindustrialisation. However, there are now global threats from the health risks associated with climate change.

Table 25 Checklist for pollution health risk case study

Causes	Impacts	Management
Pollution types/sources? Location? Point/diffuse? Incidental/sustained? Atmospheric/hydrosphere/ terrestrial? Noise? Caused by economic activity? Primary/secondary/tertiary/ quaternary? Incident or sustained pollution? Where is its location on environmental Kuznet curve?	Short term? Long term? Direct/indirect impacts? Actual threat to humans? Evidence of pollution fatigue — public outcry? Pressure group? Internet site/blog?	Who is involved? International governments? TNC/NGO/Foundation? Local pressure group? Short-term/long-term management? Type of strategy: prevention at source/treatment at sink? Strategy used, why? Policies used, why? Evidence of sustainable management?

Research activity

Research the health risks from enhanced global warming and climate change using your AS notes and references such as the 2004 *Up in Smoke* report on sub-Saharan Africa.

Compare the management of chemicals used by EU farmers to those of Green Revolution farmers such as in the Punjab, India, where there may be health consequences of long-term use of DNA.

The link between pollution, economic development and changing health risks may be focused around the **Kuznet** environmental model, based on the Western world's experiences.

Research activity

Research the Kuznet environmental model (Figure 46) and assess how this model fits rapidly changing areas like Russia, China or India. Does it link with the epidemiological model?

Contrast the management of pollution in the UK with the management in a rapidly developing country such as China. City-based data are readily available on the internet for cities like London, Mexico City, and Shanghai.

Look at international policies, such as the latest EU Directives on agricultural chemicals, water quality or electronic waste disposal. On a global level, assess the effectiveness of the UN Basel Convention that restricts transnational trade in toxic waste. Key resources: Blacksmith Institute, Greenpeace, Friends of the Earth.

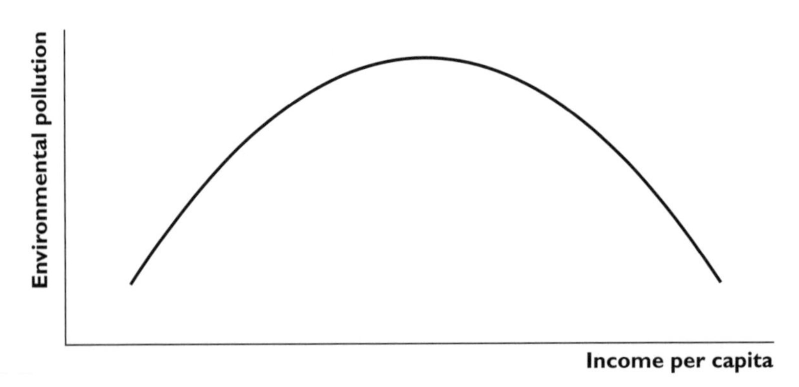

Figure 46 Kuznet's environmental curve

Enquiry question 4

How can the impacts of health risk be managed?

The socioeconomic and environmental impacts of health risk

The short and long-term impacts, or burden, of health risk, are now more complex with the so called 'health divide' becoming an increasing issue for the sustainability of our natural and socioeconomic environment. Health is a major driver of global and local economies, but the costs of healthcare are escalating because of:

- population increase, especially an ageing population
- a rise in poverty and a more vociferous middle class with higher expectations of healthcare
- technology and medical expertise — availability of expensive technology and care in prevention and treatment.
- consumer demand, increased by the media, internet information and demands for more social equity in healthcare
- rising pollution and environmental health risks from workplaces and indirectly from climate change
- global interconnectivity — globalisation of health expectations and faster movement of infectious diseases because of migration and travel. Also media-led panic about diseases such as SARS, avian influenza and swine flu

Research tip
As you collect information on your chosen health risks make sure that you include management and estimates of financial and social costs of combating the risk. There is a wealth of information on HIV/AIDS and obesity-related risks.

How health risk impacts have led to differing management strategies

Some health risks and environments are harder to manage than others (Figure 47). Health risk cannot be tackled on its own since it relies on the complex socioeconomic structures of any society.

HARDER?	• Longer-term health risks: often chronic diseases, e.g. depression, obesity and diabetes • Also incurable infectious diseases, e.g. HIV/AIDS	• Shorter-term health shocks: mental and physical traumas associated with disasters, plus infectious diseases resulting from breakdown in built environment (loss of water, sewage, housing, food supply) • Longer term relief as seen in 2008 in the Burma floods, and refugee camps in Democratic Republic of Congo	EASIER?

Figure 47 Spectrum of difficulties in managing health risks

Public health intervention includes surveillance, vaccination and family planning. It aims to prevent rather than treat diseases, with education a top priority. Healthcare involves prevention, treatment, and management of disease for individuals and communities by medical, nursing and associated health specialists. Prevention is preferable, but often more difficult to achieve than treatment, although simple schemes can produce positive results, e.g. mosquito nets preventing transmission of malaria.

Faster and more publicised types of health risk (such as HIV/AIDS, SARS) get higher priority than, for example, mental illness.

Latest concerns centre on obesity-related diseases, partly due to the globalised **food transition** involving higher fat and sugar intake. By 2015, 2.3 billion adults may be overweight and over 700 million obese (WHO).

Well-established **health systems**, such as in the UK, aim to promote, restore or maintain health. They have an integrated set of facilities and personnel, and have a hierarchy of primary, secondary and tertiary care. They have evolved from informal, small-scale, often family-based systems, into large, often government-run systems. There has also been a rise in healthcare provision by private companies and NGOs, e.g. BUPA, Red Cross.

Research activity
Contrast the healthcare system of the UK with that of China, or the USA. All are struggling at different scales and in different ways to reduce risks.

Find out the range of agencies responsible for healthcare in your own community.

Growth of international efforts to tackle health risks is linked to:
- increasing scale and issues involved
- globalisation and interconnectivity of world economies, plus political and financial linkages

International **minimum** and **ambient standards** are now common, as are flows of money, personnel and technology across the world. These are funded by international institutions like the United Nations or World Bank, or philanthropic NGOs ranging from Oxfam to the Bill and Melinda Gates Foundation. However, whether policies can be implemented effectively is still up to individual nations and individuals.

Differences in the effectiveness of health risk management, and the role of sustainability

The largest international health-risk initiative to date is the 2000 UN Millennium Declaration, adopted by 189 countries, which created a '**roadmap**' for **health sustainability**. Health is linked directly to three of the eight Millennium Development Goals, and contributes indirectly to all others. (See: **http://go.worldbank.org/RJWW34 HS90**) Hunger and malnutrition were not included in the goals, but are essential to health quality. Reduction in diseases like HIV/AIDS, malaria and tuberculosis is expected, but most countries are currently off track, especially sub-Saharan African countries and India (World Bank 2008). This is despite the doubling of health-related aid in 2000–06 from a whole range of players.

Research tip

There is no doubt that today China presents one of the largest challenges to combat pollution and reduce health risks and is worthy of closer research, especially on how its latest 'Healthy China 2020' plan is likely to succeed. You need to set up criteria to assess effectiveness, such as capital and running costs. Also, involvement of local community and effects on health (use indicators like death rates). Key references: WHO report on disparities in health risk China: **www.who.int/macrohealth/action/CMH_China.pdf** and *New Scientist* October 2008 on China's health system: **www.newscientist.com/article/dn14986-worlds-largest-health-system-rejects-free-market.html**

The most important single global advisory organisation is WHO, which has projections for future health scenarios, and an overall goal of reaching an 'age of sustaining health' (Figure 48).

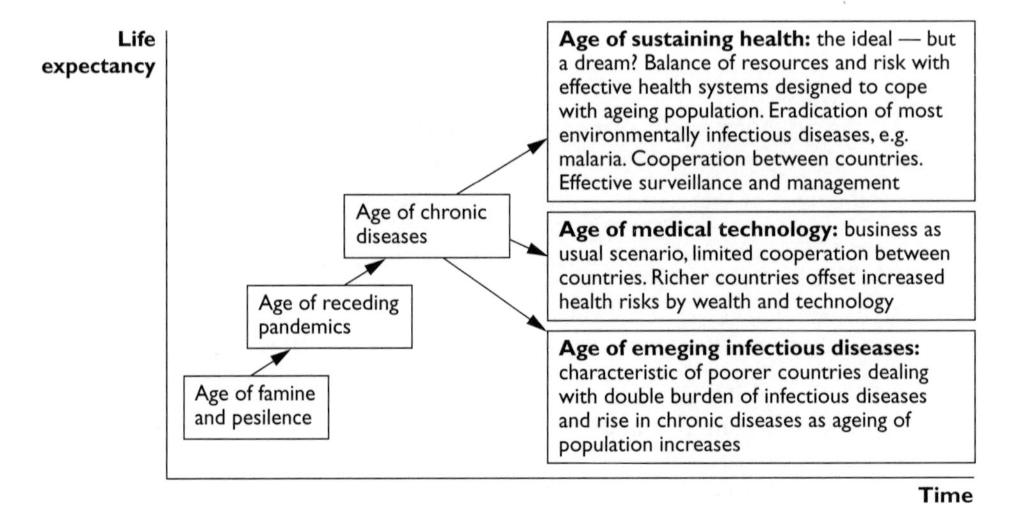

Figure 48 Health futures: WHO health transition model

Consuming the rural landscape: leisure and tourism

Introduction

Leisure is a broad term that traditionally means non-work time. **Tourism** is usually taken to mean the temporary or short-term movement of people to destinations outside places where they normally live and work, and their activities during their stay at these destinations. Figure 49 shows the relationship between leisure, recreation and tourism, and Figure 50 illustrates the range of leisure activities.

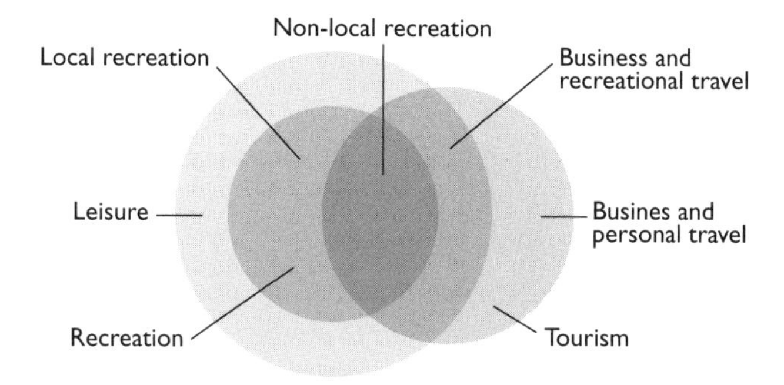

Figure 49 Relationship between leisure, recreation and tourism

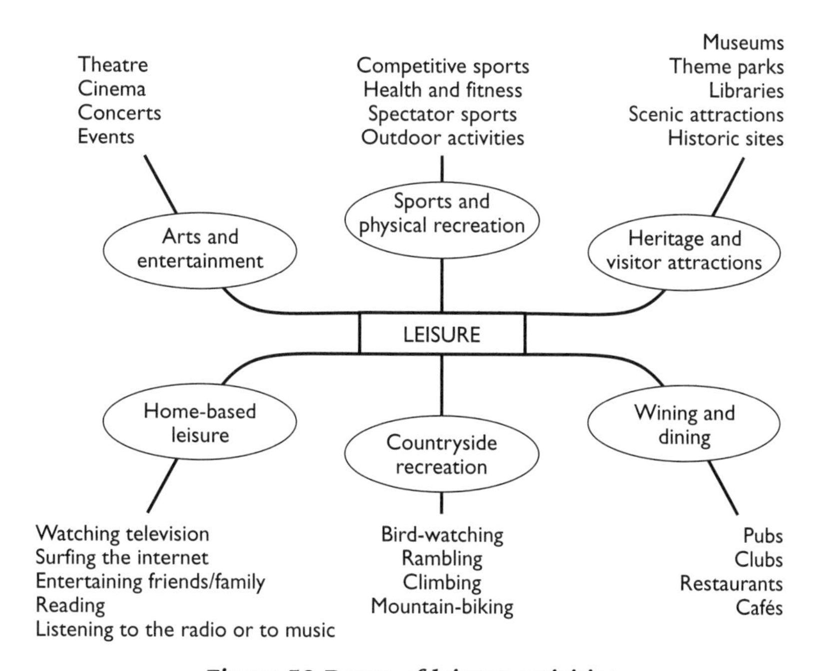

Figure 50 Range of leisure activities

Tourism is economically important — for example, the tourism industry in the UK is worth around £85 billion, more than 4.5% of the UK's GDP (2009 data). There have been three main drivers that have led to the commercialisation (or 'consumption') of many rural landscapes:

- **Economic changes** such as falling farm incomes have meant that new uses must sometimes be found for rural land if local communities are to gain sustainable incomes.
- **Social changes** have left many sectors of people finding that they have greater time available to devote to leisure and tourism activities.
- There has been a **cultural shift** — people's view of the countryside has changed and it has become something to cherish and defend, rather than simply exploit for food and minerals.

Post-productionism is the change in farming, moving from maximum food production to more sustainable agriculture and diversification into leisure activities in rural areas. In particular there has been increased emphasis on more active pursuits such as cycling and walking, together with utilisation of the cultural (including food) aspects of the countryside. This is in contrast to activities that were popular 50 years ago such as 'going for a drive' or 'picnicking'.

A more intensive exploitation of the countryside for leisure has not come without costs. There has been increased pressure on often fragile landscapes, together with displacement of local traditions and communities. Conflicts are increasingly common. All these issues require careful analysis and management to find solutions that are economically, socially and environmentally sustainable.

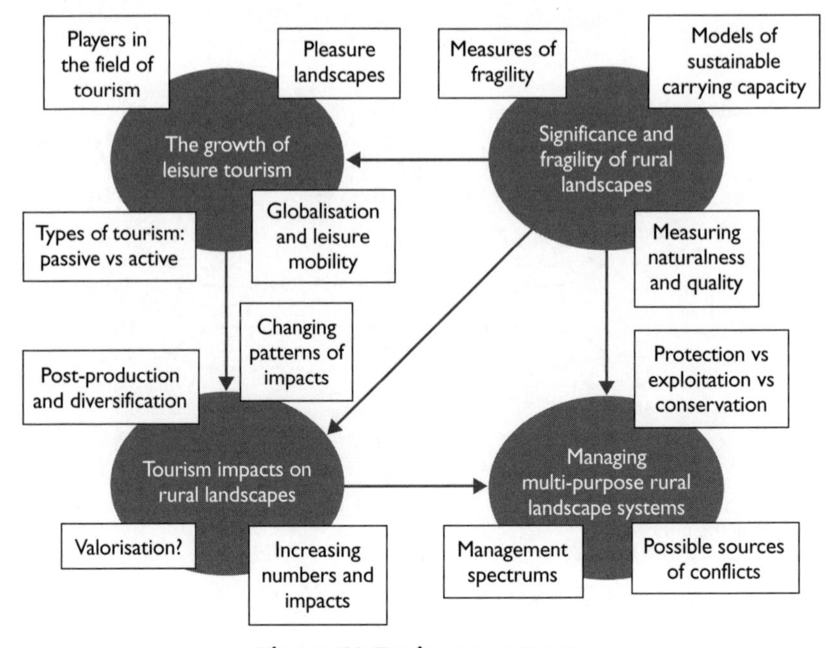

Figure 51 Topic concept map

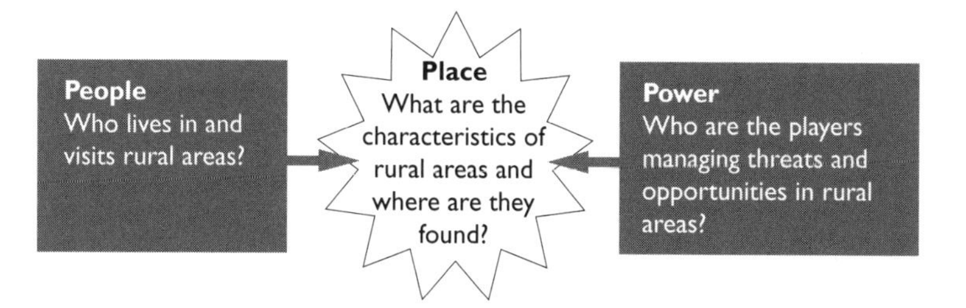

Figure 52 The synoptic context

Table 26 Checklist for consuming the rural landscape

Enquiry question 1 What is the relationship between the growth of leisure and tourism and rural landscape use?	Enquiry question 2 What is the significance of some rural landscapes used for leisure and tourism?	Enquiry question 3 What impact does leisure and tourism have on rural landscapes?	Enquiry question 4 How can rural landscapes used for leisure and tourism be managed?
• Active vs passive pursuits • Reasons for changes over time — becoming more diverse • Leisure vs tourism • Different groups and players — their roles and responsibilities • Possible sources of conflicts and between whom • Values and attitudes of different users and beneficiaries	• Landscape values (remote, accessible, wilderness etc.) • Ecological vs physical (utility) values • Notion of fragility and sensitivity • Wilderness concept • Qualitative vs quantitative environmental measures • Sustainable carrying capacity	• Negative impacts of excessive use: trampling, pollution, erosion, disturbance etc. • Positive impacts of tourism: conservation, increased awareness, protection of heritage sites • Changing impacts over time (increasing or decreasing?) • Comparing threats and opportunities in areas with different levels of development	• Arguments for and against management of rural landscapes • Developing a 'conservation spectrum' • Players involved in conservation — choices and conflicts • Evaluating the effectiveness of different approaches and options
Key refs used	Key refs used	Key refs used	Key refs used

Research activity

Use Table 26 as an initial stimulus and carry out your own research audit to obtain background sources of information. Make a note of any key references used.

Enquiry question 1

What is the relationship between the growth of leisure and tourism and rural landscape use?

The consumption based economy

One of the most significant changes in the countryside has been the transition from an economy based on food production to one based on consumption (hence 'consuming' in the title of this option). This consumption based economy is diverse and includes things like financial services as well as the more obvious leisure and tourism activities. Figure 53 shows different types of leisure that are associated with particular locations along a spectrum from rural–urban fringe to remote wilderness areas.

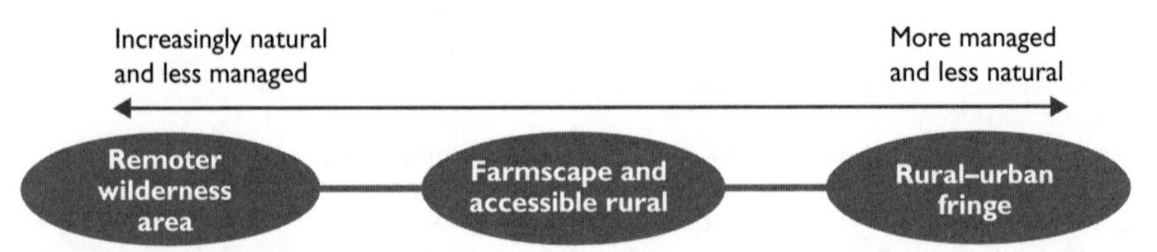

Wilderness areas are heavily protected against development so leisure and tourism is generally restricted and highly managed. Many locations remain inaccessible and off the beaten track – they are exclusive. Landscape quality is high and often well protected by legislation. Typical activities could include:

- wild camping and trails
- tourist trails
- wildlife spectacles (low key)/tourist gaze, e.g. Antarctica
- engaging with local culture, e.g. Masai village, Kenya
- sustainable ecotourism in rainforests

More accessible rural areas may be dominated by a post-production agricultural landscape, together with agri-environmental schemes. Large numbers of tourists can be accommodated in honey-pot areas. Car parking and visitor facilities may be common. Examples may include:

- organic farms with visitor centre/speciality food shop
- tourist accommodation and holiday lets
- craft centres
- wildlife/'touchy-feely' farms
- paint-balling and quad biking
- woodland trails (e.g. mountain biking)

Highly accessible rural landscapes (typically rural–urban fringe or urban hinter-land) are often battlegrounds between conservationists, planners and developers. Conflicts may occur over both activities and development, especially near to greenbelts or other specially designated areas:

- horse stabling and livery
- golf courses
- craft shops/farm shops
- garden centres/horticulture
- barns for small rural enterprise
- organic food-box schemes
- fish (trout) farming
- dog breeding
- vineyards

Figure 53 Different types of leisure and their location

Research activity

Rural tourism encompasses a wide range of activities. Research these contrasting types of activity:

- passive or 'traditional activities', e.g. fishing, walking, driving
- active or 'new activities', e.g. paint-balling, quad biking
- ecotourism and elite tourism compared to mass tourism

How and why are these activities different and how did they emerge? Where do activities such as holiday cottages and theme parks fit into this picture? Visit Tourism Concern for some introductory ideas: **www.tourismconcern.org.uk/**

An important concept linked to the development and geographical spread of tourism is the **pleasure periphery**. This concept tries to show the spread of tourism, from a point source or single location, to a much more diffuse or global activity:

- 1800: source — close to home/local
- 1900: periphery (1) based in northwest Europe
- 1930: periphery (2) extends to western Mediterranean
- 1950: periphery (3) includes all of the Mediterranean
- 1970: periphery (4) long distance travel becomes more readily available
- 1990: periphery (5) tourists are able access the world's remotest places

The growth of tourism and leisure can be linked to several key drivers:

- increasing family wealth and income
- more spare time for leisure
- increased holiday (paid)
- greater personal mobility
- greater awareness of far-off places through media, e.g. internet, books, magazines
- cheaper air travel
- rising awareness of extreme sports

The range of landscapes

There is no universally agreed definition of 'rural' but population statistics often form the basis for measurement. For example, in the UK, a settlement of less than 10,000 people is *not* classified as an urban area and by default is rural.

The distance a landscape or activity is from a large centre of population can have a significant impact on its characteristics. Some geographers classify 'wilderness' as being the most extreme rural landscape (see below), while the rural–urban fringe is the most accessible.

Concept of wilderness

Wilderness areas are remote parts of the world whose unspoilt characteristics have ecological, scientific and/or cultural and aesthetic value. Wilderness areas can be big (e.g. continental areas of the Americas) or smaller (e.g. corners of rural Britain — such as the granite tors of Dartmoor).

Wilderness landscapes are commonly perceived to be those that have avoided being modified by any significant human impact. However, many landscapes perceived as 'natural' have been modified by people. For example, the UK's 'wild' moors were historically cleared of all trees to make way for sheep farming and grouse shooting.

Find out more at The Wild Foundation: **www.wild.org**

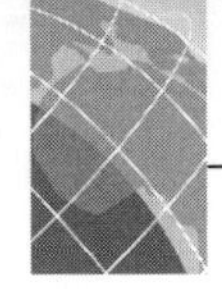

Research activity
Copy and complete the table below which looks at the different players involved in rural landscapes and their possible attitudes towards development. With arrows (last column) link together any possible sources of conflicts.

Leisure and tourism players	Examples	What they do and attitudes to rural landscapes	Possible conflicts (link with arrows)
Governments		May provide funding/legislation to promote diversification, development, etc. Likely to have social or economic benefits	
Intergovernmental agencies			
NGOs and pressure groups	Greenpeace, Wildlife Trusts		
Local/regional authorities	Natural England	May help in the promotion of rural landscapes to diversify economy, e.g. 'surfing-tourism' in southwest England. Can potentially create hot spots and therefore traffic issues, pollution, congestion, etc.	
Communities			
Farmers and local businesses			
Individuals			

Enquiry question 2

What is the significance of some rural landscapes used for leisure and tourism?

The value of rural landscapes cannot be easily defined in terms of monetary worth. Historical and cultural resources are intrinsic components of the landscape, but difficult to price. The ecological and physical value of landscape are particularly significant and again it is problematic to assign a monetary value to these. So landscape value should be regarded as having three components:

1. **Ecological and physical value** — the condition, attributes, and content/resources of a landscape.
2. **Sociocultural value** — how it contributes to physical and mental health, learning, etc.

3 **Economic value** — as quantified by direct market value, e.g. land price if it were sold.

Many rural landscapes that have high ecological value may also be fragile. Fragility refers to how stable a landscape is, or how easily it can be damaged. To assess the sensitivity and fragility of a landscape's character decisions have to be made with regard to:

- liability/potential loss of a landscape's key attributes and characteristics
- whether or not the landscape could be restored (this may be particularly important for ancient landscapes, e.g. woodlands pre-1600 or old monuments)
- whether important visual or aesthetic aspects of its character would be likely to change

Consideration should be given to the introduction of new elements that may have a significant impact on landscape character. This could include new buildings and infrastructure such as road, rail, airports, power lines, data cables and pipelines, as well as invasive non-native species of plants and animals. Figure 54 gives a more detailed example of how landscape sensitivity and fragility can be determined.

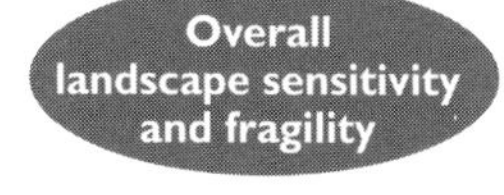

=

Landscape character sensitivity
Based on judgements about sensitivity of:

Natural factors
Vegetation types
Tree cover type/pattern
Extent and pattern of semi-natural habitat

Cultural factors
Land use
Settlement pattern
Field boundaries
Enclosure pattern
Time depth/historical significance

Landscape quality/condition
Intactness
Representation of typical character
State of repair of individual elements

Aesthetic factors
Scale
Enclosure
Diversity
Texture
Pattern
Colour
Form/line
Balance
Movement

+

Visual sensitivity
General visibility
Landform influences
Tree and woodland cover

Population
Numbers and types of residents
Numbers and types of visitors

Mitigation potential
Scope for mitigating potential visual impacts

Figure 54 Factors to consider when judging overall landscape sensitivity

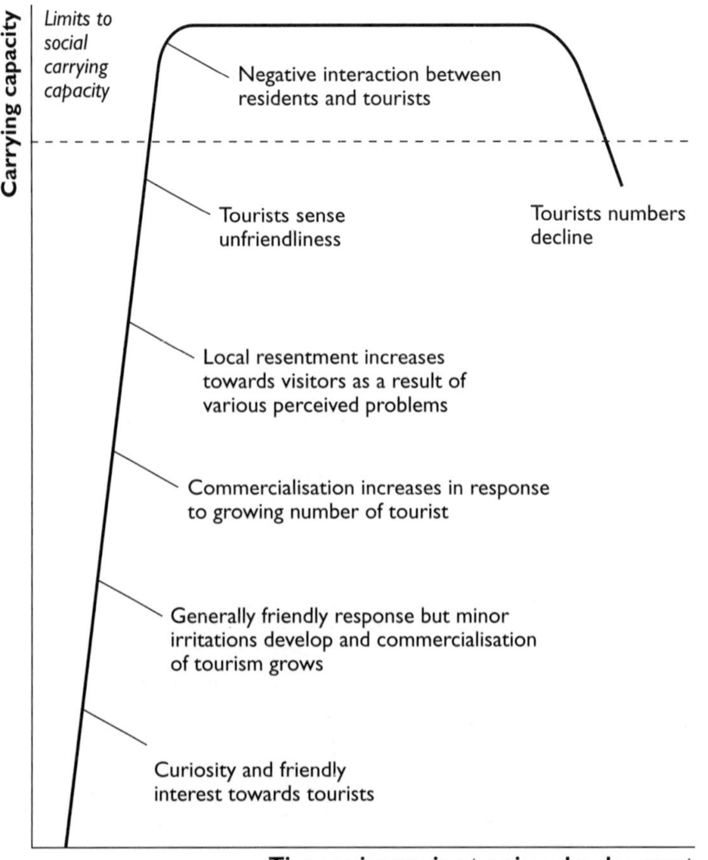

Figure 55 Social carrying capacity

Research activity
Carrying capacity is an important idea. Find out more about the different types including (1) physical carrying capacity, (2) ecological carrying capacity, (3) social/psychological carrying capacity (see Figure 55) and (4) economic carrying capacity. These may be particularly important when set against a background of access capacity, commercial capacity, construction capacity, service capacity and transport capacity. For active tourism see: **www.active-tourism.com/DefinitionActTour.html**

Quantitative and qualitative methods can be used to measure landscape quality and ecological value:

- **Quantitative** means numerical data — for example a species biodiversity count or a species frequency, e.g. using a transect, or volume of timber in an area or an actual land value price. An advantage of this method is that data can be readily processed using statistical approaches. Numerical values can also be used to provide an array of data for GIS visual processing systems.

- **Qualitative** means subjective or non-numerical data. This may include the overall 'look' or 'feel' of the landscape and could be assessed using an environmental quality index, field sketches or photographs. However, these data can be subjective and based on interpretations made by researchers as they explore their surrounding environment. Often qualitative judgements are considered unreliable.

Enquiry question 3

What impact does leisure and tourism have on rural landscapes?

Positive and negative impacts

Impacts of people on the rural landscape can be positive as well as negative. Over the last 200 years, legislation has been passed to grant animals, plants, ancient monuments, etc. protection from harm or exploitation. Important benchmarks include the formation of the RSPCA (1824) and RSPB (1899), IUCN (1973) and the establishment of World Heritage sites (1972).

Research activity

Track the changes over time that have occurred in an area of rural landscape that you are familiar with. Do some historical research to find out about past land uses and current operations. Try to be evaluative — i.e. look at the balance between the changes and whether they have provided advantages or disadvantages.

An example of a change in the rural landscape that has provided much debate is the continued development and expansion of golf courses.

Popular (especially honey pot) sites often experience vegetation trampling close to footpaths. Air, land and water pollution may also blight rural landscapes, even remote wilderness regions. For example, the beaches of Antarctica and Alaska are strewn with the plastic flotsam that has been lost overboard from commercial shipping. Areas that are more accessible will usually receive more tourists.

While it is generally assumed that more visitors mean more landscape impacts, this is not always the case. High impact sites often have greater regulation to control visitor movements and activity patterns. In contrast, a small number of visitors in unregulated or wilderness environments can do considerable damage by felling trees for firewood and dropping litter.

Emphasis on negative environmental impacts can obscure the fact that tourists can bring benefits. Any research and discussion of the impacts of tourism needs to be balanced and objective, rather than a 'tale of woes'. It is relatively easy to find examples of tourism revenue generating benefits including better wildlife protection and conservation, restoration of historic buildings, footpath repairs and even improvement of sanitation following the development of tourist hotels.

Research activity
Use the table below as a structure for assessing the positive and negative effects of leisure and tourism. You should use examples to support your ideas. How and why to the threats or opportunities associated with rural tourism vary according to the level of development of the country/region?

	Positives	**Negatives**
Social		
Environmental/ physical	Restoration of historic buildings, upgrading of resorts, improvement of polluted environments...	Soil erosion (especially following deforestation), depletion of local resources, damaging construction traffic, visual pollution...
Economic		

Commidification of the rural landscape

The promotion of local produce and traditional or specialised food products can be an important component of the valorisation or commidification of the rural landscape. Here there is emphasis on direct sales to consumers by farmers and local producers. Moreover, the sites of local food production, such as farms, dairies, vineyards and breweries are additionally marketed as tourist attractions, creating a second income stream and reason to visit.

Another way in which a rural area may market or re-brand itself is through the use of film and new media. Table 27 gives examples of a range of films that have helped promote the rural landscape.

Table 27 Films and television programmes that have promoted the rural landscape

UK locations	
Harry Potter series (2001–present)	Northumberland (Alnwick)
Pride and Prejudice (2005)	The Peak District
The Wicker Man (1973)	Dumfries and Galloway
International rural locations	
Passage to India (1984)	Marabar Caves, nr Chandrapore, rural India
Narnia series (2005–present)	Poland (Table Mountains) and New Zealand
Lord of the Rings series (2001–03)	New Zealand (Kaitoke Regional Park)
The Beach (2000)	Thailand (Phi Phi Islands)
Star Wars (1977)	Tunisia (Chott el Djerid)
Brokeback Mountain (2005)	Canada (Alberta)

Television also establishes honey pots for rural tourism. *Balamory* is a children's show that began filming in 2002 on the Scottish island of Mull, which subsequently experienced an influx of tiny tourists (accompanied by their parents). *Heartbeat* (Yorkshire) — rural set locations have become popular with tourists.

Enquiry question 4

How can rural landscapes used for leisure and tourism be managed?

Many different groups of people and organisations (players) are important in controlling the relationship between rural landscapes and varying forms of tourism and leisure. Some of these organisations hold direct power through access to funding, while others exercise a broader influencing or advisory role.

Rural areas are sometimes viewed by politicians as 'problem regions'. This is because they are typically poorer (more limited employment opportunities) and less accessible that their urban counterparts.

When considering the management of rural areas there are a series of questions that must be evaluated:

- What rights should indigenous people/landowners hold over their land and its ecosystems?
- Should endemic/localised species (including those that are dangerous) be restored to rural landscapes?
- To what extent should rural landscapes be restored, wherever possible, to an 'authentic' or original state?

Conservationists and land managers need to be clear about what they are trying achieve and what outcome they require (Figure 56). In the UK, sites are often micro-managed or preserved for a particular habitat type, e.g. rich grassland, ancient woodland or heather moorland.

There is a new EU mandate that establishes a duty for all member states to re-establish their native flora and fauna. In the UK this will mean the re-introduction of animals such as beavers, wolves, red kites and sea eagles.

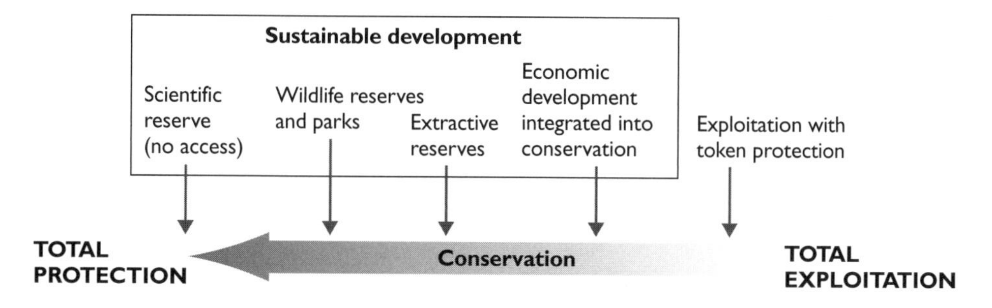

Figure 56 Spectrum of ways in which rural landscapes can be managed

There are various approaches to management. Table 28 provides some examples that can be applied in rural areas.

Table 28 Approaches to management of rural landscapes

Direct	**Indirect**
Often used in most fragile areas or in a potentially dangerous situation, e.g. waterfall, crumbling ruin. • Most time consuming and expensive • May need to start with this in short term to protect, and then move to more indirect means as education kicks in • Regulations that may entail enforcement, restricting activities or rationing use	Usually more successful in remoter locations and cheaper. • Seeks to affect behaviour through education, information and persuasion • Visitors can be informed about the impacts connected with a certain activity, or given information that encourages the use of certain areas over threatened areas • Physical alterations, such as the redirection of a trail to a more resilient area of a forest, that influence the movement of visitors
Hard	**Soft**
Paths, fences, vegetation clearance, reseeding...	Land use zoning, litter bins, interpretation signs and centres, nature trails...

Research activity

Research how National Parks are managed; identify the key players and stakeholders, and consider what are the possible sources of conflicts. Key reference — the UK National Parks website: **www.nationalparks.gov.uk/**

Questions & Answers

In this section there are six questions and responses — one for each option. You should browse through all the responses even though you will only study one option — you will find many transferable tips for the essay-style report.

The questions and research focus are similar in style to those given in the real examination. Remember that each question is worth a maximum of 70 marks.

Sample student answers are provided for each question. They illustrate a mix of A-grade and C-grade responses. The answers provided are not complete essays but are extracts from typical responses, e.g. the introduction, main body or the conclusion. Each answer is followed by examiner's comments (indicated by the *e* icon), which explain where credit is due and where weaknesses or errors occur.

Before each answer is a section that helps 'unpick' the research focus. This highlights the key areas in which the options will be assessed. It includes suggested issues for you to cover, and help on possible case studies.

In the examination, answers will be marked using the generic mark scheme summarised below.

		Marks
D	Introducing, defining, focus	10
R	Research and methodology	15
A	Analysis, application and understanding	20
C	Conclusions and evaluation	15
Q	Quality of written expression and sourcing	10
TOTAL		70

Tectonic activity and hazards

Discuss the relationship between the nature of tectonic hazards and human responses to them.

Pre-release research focus

- **Explore the causes of different types of tectonic hazard and the spectrum of responses to them.**
- **Research a range of examples of different hazard events and the reasons why responses to them differ.**

Unpicking the research focus

1. This question covers several enquiry questions from the specification:
   - Enquiry question 1: **causes** of hazards and **different events** — suggest specific examples of earthquakes, volcanoes and tsunamis and their nature, i.e. general characteristics. The event profile may help here.
   - Enquiry question 4: **responses** — the word **spectrum** means across varying responses (do nothing, adjust, or leave).
   - Response will also partly depend on **impact**, so enquiry questions 2 and 3 on impacts must be included too.
2. To narrow down revision, it may help to develop a grid like Table 1 to help you think through the various models and case studies you could choose.

Table 1 Models and case studies

Type of event	Do nothing	Adjust: Modify event	Adjust: Modify vulnerability	Adjust: Modify loss burden	Leave permanently or temporarily
Earthquake		Experimental in oil fields, Colorado	Aseismic buildings & retrofitting high & low tech. Prediction	Insurance, e.g. California Aid, e.g. Sichuan 2008	
Volcano	Nevada del Ruiz 1985	Mt Etna 1992 Operation Volcano Buster lava blasting	Monitoring e.g. Ruapehu, N Zealand Zoning of land uses, exclusion zones, e.g. Montserrat & Mt Rainier lahars	IDNDR & Hygo Framework of UN global blueprint for disaster risk reduction 2005+	Temporary evacuation: Mt Merapi 2006 Mt Chaiten 2008 Long-term: residence flanks of Vesuvius — holiday homes only
Tsunami			Protect & replant mangroves e.g. Malaysia post 2004 Knowledge e.g. retreating sea Warning systems, e.g. NE Atlantic & Pacific	Aid to, e.g. SE Asia	

3 Create some diagrams to identify reasons for different responses (e.g. Figure 1).

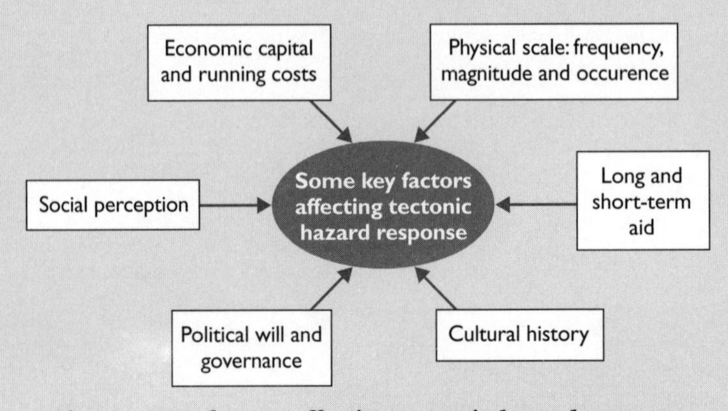

Figure 1 Key factors affecting tectonic hazard response

4 Check out some easily drawn diagrams to support your report: as there is a focus on the cause of a hazard, you should be able to draw simple cross-section diagrams as well as simple models, for example Figure 2.

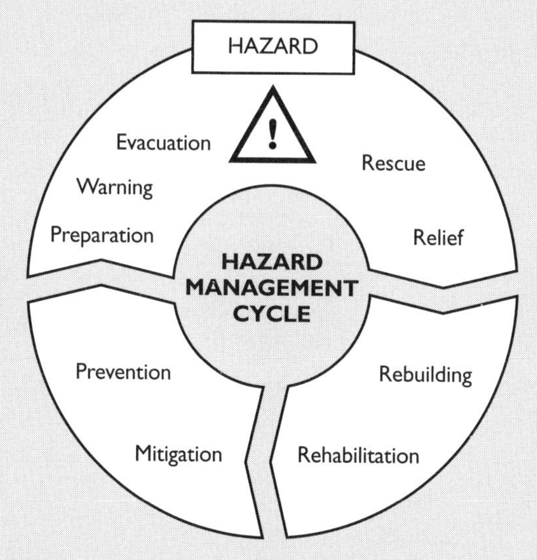

Figure 2 Hazard response and management cycle

5 Practise quick definitions with sources, for example:

From the UNISDR website:

- A tectonic hazard is a potentially damaging geophysical event…that may cause the loss of life or injury, property damage, social and economic disruption or environmental degradation.
- A disaster is a function of the risk process. It results from the combination of hazards, conditions of vulnerability and insufficient capacity or measures to reduce the potential negative consequences of risk.

For this option we will look at two example introductions.

■ ■ ■

Candidates' answers

Candidate A

Plan

- Intro: definitions, e.g. hazard
- Middle bit: examples — Mt Merapi, Asian tsunami 2004, earthquake Loma Prieta and Sichuan
- Conclusion: sum up

Introduction

A hazard can be defined as an event or process that has the potential or causes damage to people and their settlements. Natural hazards are physical processes that cause this damage, and can be tectonic, geomorphic or climatic. People who live in areas that experience natural hazards are at risk (vulnerable). The realisation of a natural hazard is a disaster.

The causes of some hazards can be prevented, although this is not usual. Other hazards can be predicted. Some hazards cannot be predicted or prevented so the only management option is to try to mitigate their effects.

In general hazards can be managed in one of three ways as shown the diagram below.

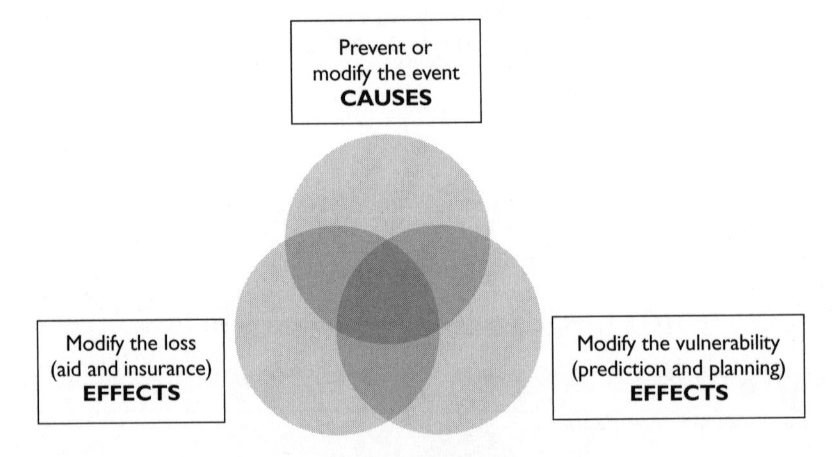

Hazard responses

The diagram above will be referred to throughout this report. Developed countries (such as the USA and Japan) are wealthy, have secure housing, emergency services, media systems and planning bodies. Less developed countries such as India and Indonesia lack this wealth. Levels of social development (literacy, healthcare) are often lower in the developing world and this creates population vulnerable to hazards. This report will focus on the contrast between developed and developing countries in terms of mitigating the impacts of hazards.

This introduction is likely to be awarded a lower middle level of marks. It is a poor plan; there is no real dissection of the question, just a vague list of examples.

There is some framework based on concepts, but no mention of case studies to be used. The focus is likely to be narrow in the rest of the report because the candidate plans to have a simplistic split between developed and developing economies. This is an outdated concept and only one of the factors affecting response, although wealth may be important as an underlying factor in response.

Definitions of key terms are incomplete — for example, tectonic hazards are not specifically introduced.

Candidate B

Plan

Relationship — graph

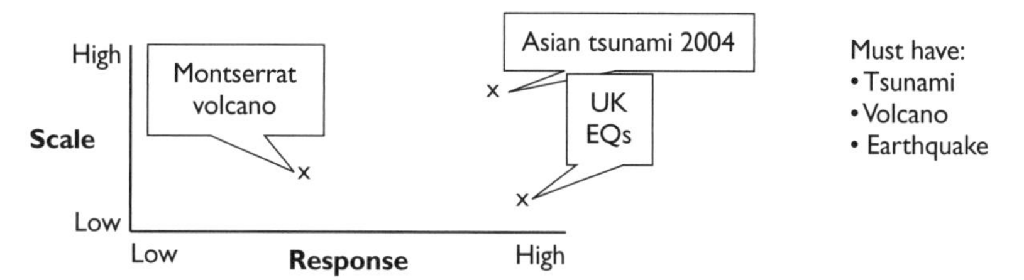

Nature = type, scale, frequency. **Other factors** = vulnerability: do risk equation **Response** = players, people.
Defs = hazard UN & model

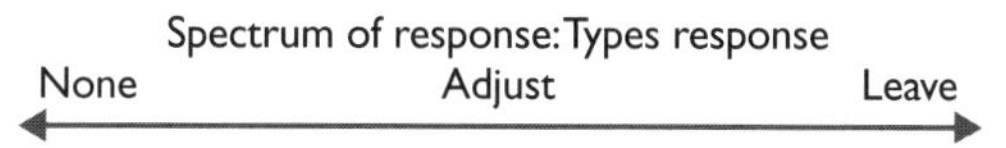

Concl — return c/s, complexity, role UNISDR in response
Vocab: aseismic, low/high tech, Benioff, mitigation, salience, pyroclastic

Introduction

A tectonic hazard is, according to the UNISDR website, a 'potentially damaging geophysical event...that may cause the loss of life or injury, property damage, social and economic disruption or environmental degradation.'

Tectonic hazards are caused by physical processes on the Earth's surface or within the Earth and their very nature in scale, occurrence, frequency and lack of prediction means they are unlikely to be prevented easily. The main focus of response is to reduce vulnerability including *mitigation*. This is shown in the table below.

Tectonic hazards

Prevented	Predicted	Neither prevented nor predicted
Lava flows — defences can stop some flows *Lahars* — channels can divert and contain them	Volcanoes — can be monitored and eruption time estimated, often with some accuracy Earthquakes are more difficult Tsunamis, as a *secondary hazard* to an earthquake, may be monitored and since they travel long distances, warnings may be given	Earthquakes — cannot currently be predicted in terms of time, although certain zones are known to be more hazardous, e.g. *subduction and transform faults* *Pyroclastic flows* — locations very hard to predict

A disaster is a function of the risk process. It results from the combination of hazards, conditions of vulnerability and insufficient capacity or measures to reduce the potential negative consequences of risk. The *risk equation* is critical to the sort of relationship that exists between the hazard and response:

hazard = (risk × exposure × vulnerability (physical, social, economic) × response)

Responses have shifted since the end of the twentieth century from developing humanitarian response and repair post-disaster, such as in the Mexican City earthquake of 1985, to more mitigation and reduction of vulnerability. This is led by the UN's decade for disaster reduction and 2005 *Hyogo Framework*. This influential *player* in responses is partly a response to the inequalities in ability to reduce vulnerability, especially among less economically developed countries. Wealthier economies (such as the USA and Japan) have more *technological fixes* such as aseismic buildings, emergency services, media systems, literacy and effective civil and planning bodies. The responses of rebuilding and rehabilitation happen everywhere, but funding is usually better in the developed world where insurance exists.

In general, hazards can be managed in one of three ways as shown in the diagram below.

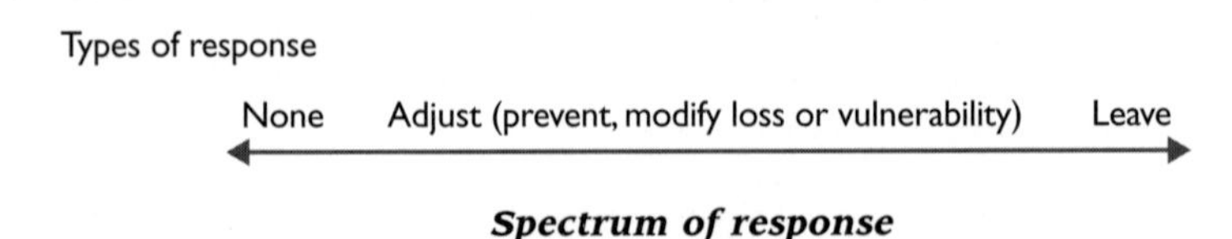

Spectrum of response

The diagram above will be referred to throughout this report with contrasting case studies of well-documented earthquakes like Izmit, Turkey (1999) as well as more recent tectonic events in Sichuan, China (2008) and L'Aquila, Italy (2009) where both a developed economy and a *transition economy* responded quickly post-event. The tsunami in the Indian Ocean (December 2004) will be used to show how the scale of the event may dictate the response, both by those directly affected and by international agencies, and the setting up of the TWS in the Pacific Ocean tries to address a need for reduction in vulnerability by giving information pre-event. Lastly volcanoes, possibly easier to predict and respond to, will be evaluated by using the evacuations of Mt Chaiten (2008) and land-use zoning of Vesuvius compared with the *do nothing approach* of Nevada del Ruiz (1985). Hazard *salience* may reduce response, however, as shown in the complacency in earthquake-prone California where many have forgotten the 'big one' predicted.

This is likely to be awarded top marks. It has a useful plan and relates to the question. It:

- develops a focus
- includes an indication of a framework, both by concepts and case studies
- gives accurate definitions of key terms, which are referenced from reputable research source. This will help the marks for researching generally

It has already started to accumulate marks for an appropriate report style with good terminology (words in italics).

Cold environments: landscapes and change

Cold environments are increasingly coming under pressure from conflicting demands. Discuss.

Pre-release research focus

- **Explore the challenges that glacial and periglacial environments are increasingly facing.**
- **Research a range of cold environments with differing uses and how effective their management is.**

Unpicking the research focus

Start by sorting out the **uses** of cold environments, for example with a simple spider diagram.

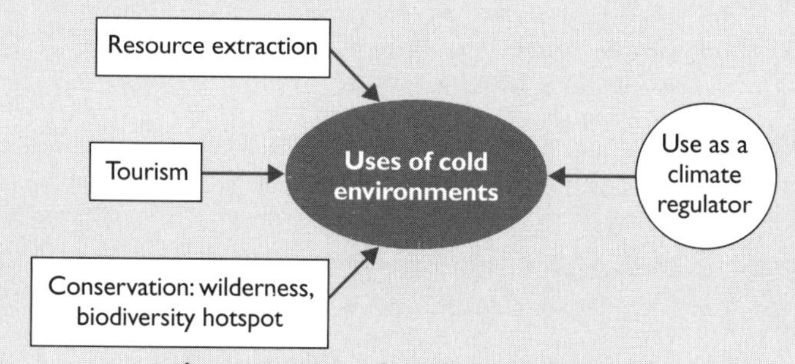

Figure 3 Uses of cold environments

Now sort out what **challenges** there are at present, in the past and in the future. The words 'increasingly facing' suggest a temporal element is involved, so you will need to find out whether the challenges have increased, decreased or changed in type over time.

A simple table such as Table 2 may help — add examples as you go.

Table 2 Challenges: cold environments

	Glacial	Periglacial
Past	Past glaciations: challenge of extreme climate conditions, people had to adapt or die, e.g. Inuit N America Marginal agriculture, low productivity, e.g. Switzerland, Lake District Use for power — HEP Antarctic Treaty	Alaskan conflict gas/oil production vs conservation Trampling, e.g. Cairngorms
Now, especially climate change	Melting, recession — Arctic, Antarctica, Himalayas, Rwenzori Mountains E Africa Post productive landscapes, increasing role of tourism — Lake District More hazards — avalanches Increased concern over loss of fragile ecosystems	Increasing melting, thermokarst Tar sand exploitation, Nenets of Russia, Trans Alaska Pipeline, Arctic National Wildlife Refuge More conservation: Cairngorms National Park
Future	Increasing tourism, yet less snow — climate change — artificial snow, new activities, e.g. extreme sports, mountain biking Increasing demands on Antarctica — World Park?	Continued

You need to look at effectiveness of management in reducing challenges. Set up some criteria, e.g. cost, longevity, involvement of local/indigenous people — even sustainability.

Ensure you are clear on the players involved: individuals to international organisations, governments to businesses, IGOs and NGOs.

Create a list of useful vocabulary: relict landform, thermokarst, indigenous, dynamic equilibrium etc.

Plan

BUG the question:

- **B**ox the command word/s
- **U**nderline the key words
- **G**lance back to ensure you understand the whole question and choose case studies/examples

Cold environments are increasingly coming under pressure from conflicting demands. Discuss.

The focus is on pressure and conflicting demands (remember the spider diagram from the pre-release stage), so sort out which case studies will be best to show a range of cold environments and a range of demands — you could do a quick table (Table 3).

Table 3 Cold environments, examples of case studies

Argument	Glacial	Periglacial
Yes — more pressure	Switzerland from HEP, tourism, changing climate	N Slope Alaska — resources — conservation
No — less pressure because of management	Antarctica so far — resources versus conservation	Cairngorms National Park Tourism — conservation

Structure

Always start with the main theme of the question — in this case *more* pressure, and then move to the opposing argument, or anomalies to the main trend. Here it is the concept of less pressure because of either reduced demands or mitigation from effective management.

You may be able to structure the report in different ways. Here are some different methods:

- by case study: safe, often pedantic and repetitive
- by type of area: glacial, periglacial, low and high latitudes
- by amount of pressure: high to low to reduced. This is more complex but may get higher marks
- by type of demands and conflicts: e.g. tourism versus conservation and then resource extraction and conservation
- by timescale: past, present, future

The following examples of student work show a conclusion to this question. Read them both and consider the examiner's comments.

■ ■ ■

Candidates' answers

Candidate A

Section 3: Final conclusion

The examples used in my report show that cold environments vary greatly in their character, and that relict glacial areas like the Lake District as well as the larger scale Antarctic glacial landscape are experiencing rising pressures. These pressures are often from conflicting demands such as more tourists, or big companies looking for resources. They can also be just from one demand, for example big HEP projects in Canada or the Himalayas. Many people want to keep cold environments as pristine as possible and stop wildlife and ecosystems being disrupted. There is the growing pressure of environmental change because of climate changes. Even periglacial areas are being affected by this, such as in the thermokarst areas of Alaska around Anchorage, where increasing temperatures are causing problems in permafrost cycles and collapsing building structures.

This conclusion does return to some of the concepts and examples used in the report, but it does not return explicitly to all the main ones used and is rather short. At the end it drifts off into general pressures rather than those from conflicting

demands on cold environments, and even adds a new case study to the conclusion (thermokarst in Alaska) without connecting it back to the main question.

This conclusion would be awarded an average grade, because it has 'meaningful comments', but only 'selective recall of content' of the report. This response also had some good ongoing evaluation (not seen in this extract).

Candidate B

Section 3: Final conclusion

In conclusion, it can be seen that all cold environments experience pressures from human activity but varying in scale and degree of directness. Some of these pressures are direct, such as in the present largest global dam construction boom of the Himalayas for HEP production for the energy-hungry transition economies of south-east Asia. The environment, both human and physical, is irrevocably changed by dams such as India and Bhutan's Tala dam project, and this causes conflicts, particularly with conservationists. More indirectly, the underlying anthropogenic causes of current climate change are raising temperatures and trigger more extreme events like glacial lake outburst floods, which in turn threaten glacial systems from Antarctica to the Arctic, and from the Himalayas to the Rwenzori Mountains of East Africa.

The Rwenzoris demonstrate the conflicting demands well since, despite being protected by UNESCO's World Heritage and local National Park status, the pressures upon them are increasing, ironically more from outside the region than within it.

This is obviously similar to Antarctica, although here, with no indigenous population, conflicting demands stem from its resource value to outside players. These include conservationists like Greenpeace and research organisations like the BAS, wealthy European and American tourists and Japanese factory fish fleets. So far the Antarctic Treaty and its various annexes have held at bay any fossil fuel resource extraction, whereas in the Arctic, BP and Statoil are examples of TNCs currently prospecting for oil and gas resources as the global energy crisis and 'peak oil' fears grow. The 30-year-old debate about drilling for oil in the Arctic National Wildlife Refuge is another example of conservation meeting exploitation head on in a periglacial environment famous for its tundra and boreal forest systems. However, the change in the main gatekeeper player here — the US government moving from Bush to Obama administration — means this cold environment may have less direct pressure.

It is obvious from this discussion that there is great complexity in the places, the people involved, and the power wielded by gatekeeper organisations. The huge variety of cold environments across the world have a common denominator: they have fragile physical systems easily disrupted by demands of whatever nature. Both active and relict glacial and periglacial environments have long been important as a water, power, mineral and tourist resource; but the added demands of the twenty-first century for conservation of physical, human and aesthetic value means increasing conflicts. The latest threat of climate change will put extra pressure and may even increase demands for water and low carbon energy production, especially on actively glaciated areas, which effectively act as a global regulator of Earth systems. There is no doubt about the increasing nature of pressures despite any mitigating management.

This conclusion would be awarded the highest level for the following reasons:

- It is clearly stated.
- It has thorough recall of content/case studies used in the report essay.
- It shows understanding of the complexity of the question, and returns to the underlying synoptic context of place, people, power. Having direct and indirect demands is a rewardable, higher level concept.

Note that the degree of ongoing evaluation within the report, which is also needed for full marks, is not obvious from this extract. You will also see that it has excellent vocabulary and is written in the third not the first person, a better style generally that will help its Quality of Written Communication mark.

Life on the margins: the food supply problem

To what extent do food security issues vary across the world?

Pre-release research focus

- **Explore why food security varies spatially.**
- **Research a range of rural and urban locations experiencing food security issues.**

Plan

Use BUG (see page 104) for a quick way of unpicking the question and setting out your main approach. For a detailed plan use the DRACQ framework (Table 4) — these letters stand for the sections in the mark scheme that examiners use to award marks:

D = discuss/define
R = research
A = application
C = conclusions
Q = quality of written communication

Table 4 Food security plan

D = 10	Definition — food security includes food supply, use FAO Large topic — say going to focus on certain aspects: famine and drought/desertification **Yes** variations — desertification. Overnutrition New York and undernutrition Mumbai **No** similarities — famine human factors politics Zimbabwe and Korea, drought Australia and Ethiopia
R = 15	Food security issues
A = 20	Basic isssues = feast–famine i.e. overnutrition and undernutrition Sub issues = increased globalisation of production and consumption Spatial: differences between urban/rural and different economic groupings: MEDC–NIC–LEDC–LDC spectrum Management — mitigation reducing issues 2009 Task Force UN
C = 15	Evaluate after each case study Return to all at end. Increasing issues globally?
Q =10	Vocabulary sections

For this option we will look at a sample introduction and a sample case study.

Candidates' answers

Candidate A

Introduction

Food security is defined as the 'reliability of food supply'. Globally an estimated 850 million people are chronically hungry and up to 2 billion people lack food security intermittently (*Edexcel* A2 *Geography*, Dunn et al.). According to the FAO there is enough food per person (using an average of about 2,700 calories of food per day). But food is not produced or distributed equally. Some areas produce much more food than others, with a range of factors involved — see the diagram below.

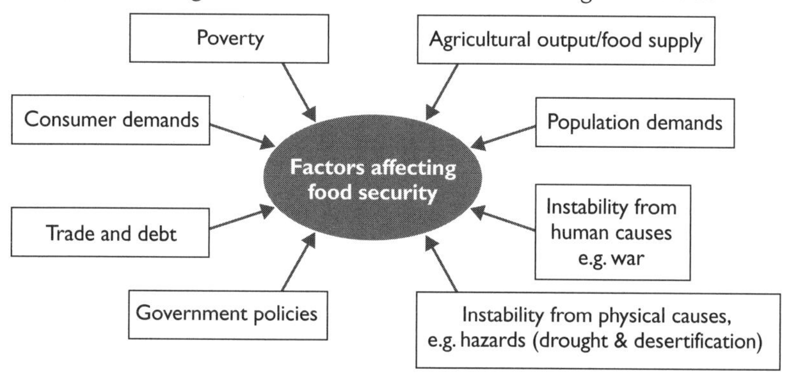

Factors affecting food security

Access to food is still seen by many as a privilege rather than a basic human right — hunger causes the deaths of about 35,000 people daily. Malnutrition is even more widespread as an issue. Surprisingly this affects richer economies too, such as Canada where 2.5 million people depend on food banks (International Development Research Centre Website), and an estimated 30 million people in the USA may be unable to buy enough food to maintain good health. So although it might be thought that only rural areas suffering desertification, like Ethiopia, have poor food security, in fact a growing issue is that of food security by the urban poor in expanding megacities such as Mumbai and even in cities in more developed economies like New York.

In recent decades, health and nutritional status have improved considerably, but are still far from satisfactory in many countries. Ironically in many richer economies the opposite issue is emerging: that of obesity — malnutrition of a different kind.

The rising middle classes of transition economies like China are generating another issue: that of changing food habits with increasingly Western preferences for meat rather than traditional types of food. This is putting pressure on rural areas hundreds of miles away in our globalised world, as soya is grown for external markets for cattle fodder rather than crops being grown for local consumption. Meanwhile, food issues in countries like the UK centre around food miles, ecological footprints, cheapness of foods, production methods like animal welfare and organic products as well as obesity.

The earliest known famine was in 3500 BC in Egypt, and famine has affected places as far apart as China, India and Ireland. The worst famine to date was probably in nineteenth century China, killing up to 13 million. The FAO states that such major disasters increased fourfold between the 1960s and 1980s, but the international donor community has become better at preventing them turning into catastrophes. Recently there have been extra efforts to address famine in places like Ethiopia by NGOs such as Comic Relief. Red Nose day was started in response to the 1985 famine there. In the LDC of Kenya in 2009 the president declared a national food shortage, showing that years of effort trying to improve food security have not been enough.

The fact that the UN set up a Task Force on Global Food Security in 2008 indicates that food security issues are far from sorted. The Task Force includes many UN agencies like WFP, WHO, UNICEF and the World Bank.

e This is a good discussion on a general range of food security issues, with relevant examples. However, it does not signpost clearly to the examiner the framework to be taken, either in terms of concepts, e.g. which food issues, or specific case studies, e.g. China, Mumbai.

In the plan there was an indication that food security issues common to different areas and different in locations would be assessed, but this simple and effective two-fold division to help argue points does not appear clearly in the introduction.

This would get a good mark but could have achieved top marks with the addition of an extra paragraph to justify and illustrate the chosen issues, as follows:

This report will focus on the following food security issues...and case studies to assess how the issues of food security vary across the world and also over time:

- the NIC megacity Mumbai because...
- national scale: RIC of China, which shows...
- current insecurity issues in Zimbabwe, especially in the capital Harare, because...

Candidate B

Factfile on China (sources: FAO, CIA, BBC)

China is a good example to show how food security varies spatially within a country, and also over time. China feeds over one-fifth of the world's population with only one-fifteenth of the world's arable land. Access to food is governed by poverty and although national economic growth is strong, it is uneven across regions. Farmers' incomes in the central and eastern regions of China continue to grow more rapidly than those in the west and southwest. Income inequality among regions, between rural and urban areas, and within regions continues to grow. More challenges of food supply are created by degraded land, water and resources, and increasing cost of labour and domestic food production. In 1958 and 1961 China had widespread famines with an estimated 15 million associated deaths. The Nobel prize winning economist and expert on famines Amartya Sen says that most famines do not result just from lower food production, but also from an inappropriate or inefficient distribution of the food, often compounded by lack of information and indeed misinformation as to the extent of the problem. In China, the reasons have been:

- largely from a series of natural disasters, floods and droughts especially, compounded by poor management
- the huge institutional and policy changes from the 'Great Leap forward ' policy involving forced collectivisation, which reduced the incentives for peasant farmers to work effectively; labourers were often forced into working in manufacturing instead of food supply

The famines were associated with land degradation, which has had more effective management to resolve it as shown in the diagram below.

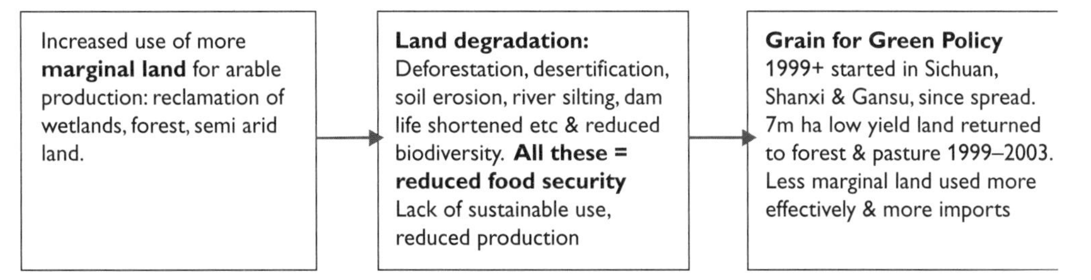

Data on China from the NGO 'Leadership for Environment and Development' established by The Rockefeller Foundation

However, despite the pessimistic forecasts made in the 1960s, famine has not been a serious issue in China in the past 50 years because of land reforms and the development of a market economy allowing wealth to spread. In fact about 20% of the 1 billion overweight or obese people in the world are Chinese and China is fast catching up with Western levels of obesity; 12% of adults in cities are now classed as obese. The speed of the nutrition transition from a low-fat to a high-fat diet is one of the highest seen globally.

Sub-conclusion: this case study shows both spatial and temporal differences in food security, and a shift from famine and land degradation as the main food issues to changing food tastes and health issues linked to obesity.

The main report section shows one well-selected case study that addresses many of the complex points involved in the question. The use of statistics and referenced data combined with an evaluative sub-conclusion ensures top marks will be awarded for selection, application and conclusions. There is also excellent geographical terminology, for example the words degradation, nutrition and transition.

The world of cultural diversity

'Cultural attitudes determine how landscapes and environments are valued.' Discuss.

Pre-release research focus

- **Explore the attitudes of different cultures to the environment and landscape.**
- **Research a range of landscapes and environments to see how these attitudes may affect their value as shown by management and use.**

Unpicking the research focus

Identify a range of different cultures and attitudes and uses to/of environment/landscape. Use the examples/case studies from Tables 5–7, which can be used or rejected in the final exam.

Table 5 Environment: different attitudes and cultures

Cultures with differing attitudes to the environment and landscape		Cultures with similar attitudes	Cultures with a change in attitudes over time
Exploitation focus USSR: Aral Sea, Chernobyl China (see changes)	**Protection focus** UK Hindu and Islamic religions Ethnoscapes: Aborigines Uluru sacred site Global: UNESCO and World Heritage Sites, e.g. Marrakech traditional market (souk) and Djemaa el Fna, Peru Macchu Picchu, Jurassic coast UK	The global green and conservation movement: a culture not bound by national frontiers e.g. Greenpeace, WWF, FoE and green political parties, e.g. German or Kenyan Green Party Global corporations/TNCs and architects and planners: cloning of urban financescapes National Trust NGO and 2009 strategy on sustainability and cultural preservation	UK — past Industrial Revolution and exploitation turned to pro environment after 1960s especially. New National Parks and Conservation Areas in cities. USA Wildscapes and ecoscapes and Obama on global warming Public attitudes to climate change Changing attitudes to wolves China possibly recently: Beijing Clearskies policy, e.g. new National Parks Rebranded landscapes e.g. coastal resort of Boscombe, Olympic site in London, cities with 24-hour culture, e.g. Birmingham

Table 6 Some types of use and management

Types of use	Types of management
No use, preservation Use for resources under surface e.g. oil, coal Use of ecological resources, e.g. forest products, water abstraction Use for urban demands: ecological footprints Use for tourism: mass, eco, elite	Top down: often government-led large schemes Bottom up: grassroots movement to preserve/protect/conserve Management from pressure groups and NGOs Sustainable management: long-term planning for future generations

Table 7 Factors affecting how landscapes and environments are managed

Social/cultural	Economic	Physical
Strength and commitment of governance Role of IGOs like UNESCO or UNEP Role of individuals, freedom and apathy Religious/ethical attitudes Consumerism and McDonaldisation	Finances available to protect/preserve/find alternatives Role of TNCs, e.g. in polluter pays NGO and Foundation funding Poverty and wealth	Scale of area or environmental problem Technology available to use Surveillance to check management working

Plan

BUG the question (see page 104 for more on BUG).

'Cultural attitudes determine how landscapes and environments are valued.' Discuss.

Candidate's answer

Introduction

There is an infinite mosaic of natural, semi-natural and human-dominated landscapes across the world, from the wildest landscape of Antarctica to the middle of downtown New York. As for the environment, this is made up of the atmosphere, hydrosphere and ecosphere as well as the land. Humans are part of the natural Earth systems, or so some viewpoints would say, since other viewpoints see humans as a separate entity with the Earth as its resource. This is the fundamental reason why there are differing attitudes, plus the fact there are so many players involved: from international ones like UNESCO and UNEP to individual governments like the UK and India, trade blocs like the EU, businesses and TNCs like Shell, green pressure groups like Greenpeace and WWF and last but not least, individual attitudes. An urbanised viewpoint is increasingly dominating attitudes, since over 50% of the population live in the megacities and world cities of the world from Beijing to London. The diagram below summarises these attitudes and the key case studies that will be used to discuss them.

Mainly exploitative attitude today	Used to be exploitative but more protective recently	Used to be protective but now more exploitative	Mainly protective attitude today
Westernised, humanist society Russia	UK National Parks and sustainability Rebranded historic/cultural landscapes, e.g. Boscombe, Bradford, Stratford, Olympics	Hindu faith, India	Romantic and ecological viewpoints GAIA concept WWF Greenpeace UK USA Environmental Agencies

Attitudes and key case studies

Section 1 An overview of attitudes with their historical and religious links

1.1 Ecological and exploitative views summary

Different people and groups attach different values to the same landscape or part of the environment. For example, trees may be valued for aesthetic, ecological, food and profit reasons depending on people's current wealth and general culture. Similarly, wolves and tigers, once seen as a danger, are now seen as key species by ecologists and an embodiment of 'wildness'. This introduces the concept of changes over time in our attitudes.

1.2 Roles of religion and technology

Long-established religions have usually viewed the environment as sacred, with every tree, river, animal and bird containing a spirit and needing reverence. Islam, Buddhism and Hindu faiths are good examples of this. The more recent, allied viewpoint is the Gaian attitude to Earth as one system. Professor Lovelock stated in 2004 that global warming is 'the response of our outraged planet'. Advocates of this view on the environment want to use technology like renewable energy and even nuclear power to manage the environmental repercussions of human-led change.

However, the attitude fostered by Christian faiths over centuries is an anthropogenic standpoint where nature is a type of utility to humans and traditionally exploited on an ever-increasing scale as technology, population pressure and consumerism have developed. Environmental degradation was so obvious by the 1960s that the Green movement was created, originally in the USA. Here, the backlash from the extreme use of the landscape, where humans were technologically dominant, combined with freedom of speech and the new perception of Earth resulting from satellite imagery and space travel, helped form some new attitudes:

- The Romantic–Preservation ethic: where nature is a type of god, vulnerable to human activity, and should be preserved or protected as much as possible. The priority use is therefore nature reserves, parks and wilderness areas — policies used by many NGOs and governments today. Unfortunately many of these can be visited only by wealthy elite tourists, usually from the capitalist culture, and those attracting most visitors may even be 'loved to death', as seen in the Lake District and Yosemite National Parks when honey pot areas become super saturated with visitors. It is an attitude that is almost voyeuristic, and any indigenous people may suffer in core areas where pristine 'nature' is designated. This has led to well-known culture conflicts, for example at the sacred site of Uluru in Australia between the native Aborigines and the government. The WWF has even expelled people from some reserves, as in Korup.

- The Resource Conservation ethic focuses on the greatest use of the resources of an area for the greatest number of people for the longest time. Exploitation of resources is reduced or not allowed. This has sustainability at its core, and is followed by most Westernised governments since the 1992 Rio Summit. This includes the concepts of Polluter Pays, Precautionary Principle, Local Agenda 21, and most recently contraction and convergence as part of tackling the pressing issues associated with climate change management. It also tries to involve all the stakeholders in an area and create a working relationship between nature and people.

1.3 Russia

However, not all cultures considered as 'Westernised' have the same change in approach to the environment as the UK, Europe and the USA. The table below is a timeline of Russian attitudes to the environment, taken from the Russian branch of Greenpeace and compiled by Tsyplyonkov: Russia has transformed itself from a

Twentieth century up to 1990	Generations of Russians have grown up with the attitude reinforced by their government, often forcibly, that the state is dominant and individuals are not responsible for environmental care. This shows in low rates of responsible waste disposal. The explosion at the Chernobyl reactor in Soviet Ukraine in 1986 was the world's worst nuclear disaster. The late twentieth century saw grandiose schemes such as diverting rivers, as in the Aral Sea tragedy, and building of many nuclear power plants including Chernobyl and polluting industries like around Lake Baikal.
1990s–2000	The huge change in political culture in the 1990s with the collapse of the Soviet Union created a short-lived change in attitude with the formation of an Environment Ministry, but by 2000 this was abolished. Separate departments mean no integrated overview of the environment exists, nor any one agency has ultimate responsibility. This contrasts with the situation in most European countries and the USA, which have overseeing Environment Agencies.
2004	Russia signed up to the Kyoto agreement in 2004 only after pressure from the EU linked with trading relationships.
2009	The basic attitude in Russia seems to be that environmental restrictions pose a threat to economic development, seen most recently in the huge environmental impacts of transforming the Black Sea resort of Sochi for the 2014 Winter Olympic Games.
2007	The task of changing individuals' perception of the environment is being led by international green groups like Greenpeace and WWF. Public outcry against a proposal to build an oil pipeline close to Lake Baikal helped persuade the government to stop its development.

Timeline of Russian attitudes to the environment

capitalist to Marxist to hybrid social type culture in the past century. Today it is infamous for continuing its exploitative attitude to the environment — both at home and in its ecological footprint: 'the Russian government is prepared to pay the ecological price for economic development'. This change in attitude is also being seen in China and India, as the following case studies of countries in economic transition show.

1.4 China and India

China has immense environmental issues, as highlighted by the name and shame website the Blacksmith Institute. However, there is the start of a shift here as the government wants to display a more developed world attitude to environmental issues, demonstrated in the 2008 Beijing Olympic clean-up and its Blue Skies policy.

In India, the original Hindu attitude to environmental protection has been lost in the drive towards economic progress, with deforestation in the Himalayas, desertification in Rajasthan, and pollution in cities, for example, pollution of the River Yamuna in Delhi and air-pollution in Chennai. The Indian environment-protection movement that opposes this degradation of natural, farmed and urban landscapes is organised largely by Westernised elites and based on Western models. It has failed to become a mass movement because it lacks spiritual foundation and content necessary to inspire Hindus (quote from the philosopher Professor David Frawley).

1.5 Zoning and mixing

The concept of *himil*, meaning protection of certain zones, has long been used by the Bedouin tribes as a custom/tradition inherited from their ancestors, and can still be seen in many Muslim countries, such as Saudi Arabia, where it is practised by the government to protect wildlife, e.g. efforts to save the Arabian oryx from extinction. Islam, as a way of life, expects humans to conserve the environment. It is not a culture but a body of principles and values, which may be assimilated into whatever the local culture is. Cultural landscapes in many larger cities across the Westernised world now feature components of such mixing: from mosques to supermarkets, and even schools such as Islamia in London, which was originally started by former pop star Cat Stevens, now known as Yusuf Islam, and is now funded by the British government.

1.6 Cityscapes

Heritage landscapes in cities are increasingly valued as a commodity, as cultural tourism helps rebrand places as different as Boscombe and Bradford. Efforts to keep individuality of places may be seen in the 'home town–clone town' debate. Globalised forms of consumerism in retail high streets and especially financescapes are often fought against to give places a special place value and resist cultural imperialism. That is why places like historic Winchester may be more valued for some elite footloose businesses than say the more utilitarian cityscape of Basingstoke, although even here the cityscape is dominated by TNC land uses like Starbucks and McDonald's. The ultimate globalised cityscape is probably the large-scale shopping mall, where attitudes to consumerist culture and domination by some large TNCs mean a very standard layout and look from Putrajaya in Malaysia to Shanghai in China, Westfield in London and of course where it all started — downtown malls in any US city.

In London the new Olympic landscape has strong overtones of environmental protection and eco-friendly/low-carbon buildings and landscapes, although the original urban landscape was partly destroyed despite local opposition in Stratford.

1.7 Subcultures

Subcultures may have an effect on landscapes too: whether rural such as the agriculturally based Mennonites in Belize or urban such as the hip-hop culture of graffiti and street culture.

Lastly, this report will discuss some of the key players other than pressure groups and governments.

1.8 Players

At the lowest scale, but by far the most numerous at over 6 billion, is the individual. Attitudes to landscape management and environmental issues vary, as shown by the 2007 HSBC survey of public attitudes towards climate change in nine countries across the world, involving 9,000 people. The HSBC Climate Confidence Index shows that people in developing economies such as China and India exhibit the greatest concern, commitment and optimism towards the problem of climate change. The greatest indifference, reluctance and fatalism is evident in developed economies, especially France, the UK and Germany. Across all countries, 68% thought governments should be playing the leading role, rather than NGOs, companies and individuals.

UNESCO is a key player in fostering cultural diversity and preserving diverse cultural landscapes globally, being the largest international organisation to date. Countries sign up to its ideals, including the UK and the EU bloc. It creates protected areas and protects buildings of global significance.

The EU has a range of policies designed to preserve and enhance the landscape and environment, from the Regional Development Fund for heritage restoration in Ebbw Vale (an ex mining and steel town in south Wales) to the LEADER and LIFE III projects (source: Europa.eu). Many of these restore landscapes, if not to an original form at least one that is less polluted and more visually pleasing, and can even become a resource: as in Swansea's 5 Parks scheme in an area once the worst degraded in Europe. This shows the ideas of Carl Sauer back in the early twentieth century, who thought that cultures develop because of the landscape but also mould the landscape as well.

At a country scale, the UK has a sustainable communities policy and many others designed to respect and improve the environment, ranging from National Parks to recycling targets, renewable energy plants to conservation areas in cities.

Conclusion

There are many complex, interrelated factors influencing attitudes to landscape and the general environment, many based on culture in its widest sense. It has been shown how religion is a major cultural influence in attitudes, but above all the role of economic development outweighs environmental costs, especially in Russia. The present backlash focusing on environmental climate change may be too late with the

tipping point described by Stern and Cox as imminent or even passed. It has been shown how international efforts are involved to both create cultural diversity and environmental protection, with NGOs like Greenpeace and IGOs including UNESCO as major players, working in conjunction with organisations like the EU and nation states. Landscapes include both wildscapes and cityscapes, and there are efforts to preserve diversity in landscapes both for philanthropic reasons and to increase their commodity value as tourism increases.

What makes this an excellent report?

- It is well referenced, from a range of reputable sources.
- It has a range of case studies, some in considerable detail, which enable an in-depth discussion.
- It covers differing scales: global, national and local.
- It uses vocabulary in a sensible, knowledgeable way.
- It relates each example/case study back to the question, with sub-conclusions at the end of each section.
- It mentions some theories effectively without getting carried away on a critique of them.
- It has a distinct report style but a flowing prose discussion.

How could it have been improved? A slightly longer conclusion, returning to most of the main points/examples would have helped. The conclusion also introduces a few new points not covered in the report.

Pollution and human health at risk

To what extent do health risk causes and patterns vary?

Pre-release research focus

- **Explore the complex causes of health risk, and any links to socioeconomic status and geographical features.**
- **Research a range of health risks and how models may help the understanding of their patterns and trends.**

Unpicking the research focus

What is a health risk? Use classification into infectious, chronic, pollution/toxic and traumas.

Causes and patterns suggests a focus on enquiry question 2: what are the causes of health risk? Other enquiry questions may help — for example, enquiry question 4 on management to mitigate causes and impacts. Start with a simple summary of a range of causes (Table 8).

Health risks include infectious, chronic and trauma risks, so you need examples of each, linked to epidemiology and health transition models, for example China, the UK.

Table 8 Range of causes

Human causes	Physical causes
Lifestyle choices: exercise and diet Vulnerability, depending on: • Wealth — employment and income: socioeconomic status • External factors of health service: state/private/insurance • Role of management and effectiveness by international/national/local authorities and NGOs and businesses and charities	Environment: geographical features — disease reservoirs, vectors Changes in environment: climate change, e.g. shift of malaria Housing: displaced people, refugee camps, e.g. Darfur Pollution: Kuznet curve suggests differences between LEDCs/NICs and MEDCs

Plan

BUG the question (see page 104 for more on BUG).

[To what extent] do health risk causes and patterns vary?

You can do some quick planning around the question as shown in Figure 4:

Have to weigh up the variations, see if any are similar, e.g. infectious disease of HIV and measles — Switzerland, USA, UK
Extent of differences — scale different in...
By end have to decide

Human causes, i.e. lifestyle choices by individuals, wealth, technology, management by WHO, Govt, NGOs, overuse antibiotics in MEDCs...

Physical causes, e.g. — local environmental conditions — river pollution, global warming

To what extent do health risk causes and patterns vary?

Show range and complexity of health risks: infectious/communicable diseases, e.g. measles, malaria
Chronic/degenerative diseases, e.g. obesity related diabetes & depression
Toxic risk from pollution — cancer villages China
Trauma — industrial accidents sweatshops

Spatial pattern:
- global, i.e. MEDC–NIC–LEDC–LDC spectrum
- local, e.g. within a city — e.g. Manchester
  Temporal pattern–epidemiological model
  Models may help even though not specifically asked for: contagious diffusion, measles and hierarchical diffusion — HIV

Figure 4 Plan

The following answer is the outline for the main analysis section, given marks in the Research and Application part of the mark scheme.

Candidate's answer

Section 2.1: Infectious diseases: the case of measles, a long-standing preventable and curable infection

Overview

According to the UN office for the coordination of Humanitarian Affairs, by 2008 the Global Measles Initiative had helped the vaccination of over 400 million children in 50 countries. The players involved are the American Red Cross, the US Centres for Disease Control and Prevention, the UN Foundation, UNICEF and WHO. They aim to reduce measles deaths by more than 68% globally and 91% in Africa compared with 2000 rates.

Case study 1: Pakistan

Pakistan is currently attempting to vaccinate 64 million children in just 1 year in the effort to eradicate this highly infectious disease, which can cause complications of pneumonia and encephalitis and a high mortality rate. Being only at stage 2 of the epidemiological model, it is still dominated by acute infectious disease, largely because it is a low-income economy (GNI of US$870). However, according to the Population Reference Bureau website, Pakistan is entering a new phase of increased chronic health risks because of its rapidly ageing population, so it will have a double burden to cope with.

Case study 2: UK
In comparison, the health risk from measles in the UK (GNI US$42,740) is minimal because of mass vaccinations in the late twentieth century. However, since 2004 there have been outbreaks, mainly because some families have chosen not to immunise their children owing to an unfounded scare over the MMR injection.

Case study 3: San Diego
The pattern of disease follows the contagious diffusion model of Cliff, where direct contact is needed between hosts of the disease. This was shown by the 2008 outbreak in San Diego USA caused by a group returning from holiday that had had contact with an infected person in Switzerland. Despite being a highly developed economy, which is really at stage 3 of the epidemiological model, Switzerland has no centrally structured vaccination plan. This particular outbreak travelled the world in less than a month because of the air travel factor. The San Diego family had refused the vaccine, and this was the first time since 1991 the city had seen measles.

Sub evaluation: The causes are basically the same, i.e. the presence of the pathogen, but it is the management that makes the health risk so different in these countries — poverty/wealth and the efficiency of the health system influencing the vulnerability of the population. The scale of the problem in Pakistan obviously needs outside assistance. Yet it comes down to individual lifestyle choices as well.

Section 2.2: The case of an emergent, preventable but incurable infective disease: HIV
Unlike measles, the newer emergent infectious disease of HIV has no cure, but with proper treatment and nutrition, mortality may be delayed. HIV is a problem for differing groups of people, i.e. its pattern differs markedly both globally and more locally.

[Here the candidate wrote a discussion of the different strains and patterns of HIV, groups of people most affected, role of antiretroviral drug campaigns, RED, Melinda and Bill Gates Foundation aid and the WHO HIV/AIDs management strategies...

Section 2.3
This was followed by an evaluation of the variation in chronic diseases, which are now creating a double burden especially in transition economies like China.

Section 2.4
And still on China, a section follows on pollution as a major cause of toxic health risk, using the Blacksmith Institute and Reuters news agency as sources. The focus is on cancer clusters like the village of Huangmenying in central Henan province, poisoned by unregulated effluent from tanneries, paper mills and an MSG factory flowing into the main water supply of the Huai River. The causes are linked with ineffective environmental quality standards and health cover in the country.

The pattern is that all age groups are affected but in a relatively small hot spot of pollution — although as a nation China is now infamous for its poor environmental quality and under-resourced healthcare system...

Section 2.5: Environmental change at a global scale: climate change and the potential rise of malaria

Discussion of past, present and future scenarios for malaria, classed as an endemic re-emergent disease in especially Africa (e.g. Nigeria), owing to a complex mix of causes: chloroquine resistance, civil unrest and migrations, increased rainfall, dam construction, reduced health budgets and drug availability, high birth rates producing vulnerable under 5 year olds...sources: Wellcome Trust, WHO, UK government DFID.]

What makes this section potentially the highest level?

- It is well referenced, from a range of reputable sources.
- It has a range of case studies, some in considerable depth, which enabled a decision on the command word in the final conclusion.
- It covers differing scales: global, national and local, although it does not manage to cover the city scale, e.g. research on Manchester's differing morbidity and mortality patterns.
- It uses vocabulary in a sensible, knowledgeable way.
- It relates each example/case study back to the question, with sub-conclusions at the end of each section, so will gain marks not just in research and application but also in the conclusions section.
- It mentions models effectively without getting carried away on a critique of them.
- It has a distinct report style but a flowing prose discussion.

How could it have been improved? A quick sketch of the diffusion model, annotated to show the examples chosen, or the Kuznet curve, which appeared in the student's research focus but was omitted in the exam, would have helped.

Consuming the rural landscape: leisure and tourism

'The use of rural landscapes for leisure and tourism is more controversial in some areas than others.' Discuss.

Pre-release research focus

- **Explore the impacts of leisure and tourism as a consumer of rural landscapes, and the role of management.**
- **Research a range of locations showing differing impacts from recreational and tourism users.**

Unpicking the research focus

Identify the key areas of the research focus — sometimes a simple bullet list is all you need:

- How leisure and tourism = a consumer.
- **The users.** Who is involved in the use and management of rural areas — people/players: stakeholders, gatekeepers within area and outside — increasing role of international organisations from pressure groups, conservation to TNCs exploiting rural areas and UN, plus individuals.
- **The uses.** What are the uses: types of leisure and tourism, including comparison of overlaps between leisure and tourism activities.
- Elite, eco, mass.
- Active, passive tourism etc.
- **Uses and impacts**: positive, negative, direct, indirect, short and long term, economic/social/environmental etc.
- Role of management in increasing/decreasing impacts.

Create a series of factfiles on places showing these different impacts. You could categorise these into a spectrum diagram (Figure 5).

Low impact	Low impact at present	Was high now lower because of effective management	High impact at present	High impact
	Antarctica Wolong S Downs	Macchu Picchu New Forest Mallorca	2nd homes Snowdonia, Caribbean e.g. Barbados	

Figure 5 Spectrum diagram

Plan

BUG the question (see page 104 for more on BUG).

'The use of rural landscapes for leisure and tourism is more controversial in some areas than others.' Discuss.

Sort out your arguments — a simple table (Table 9) can help.

Table 9 Controversial or not?

Yes — very controversial because	No — less controversial because
High negative impact — Snowdonia second homes, loss of farmland and tropical rainforest St Lucia Changing leisure use e.g. mountain bikes Glen Shee Cairngorms Increasing impact — Antarctica, Kilimanjaro Less effective management — golf courses Vietnam	Less impact — remote Better management — core zone Yosemite, Annapurna — shows TEMPORAL CHANGE Partial management — fox/drag hunting Hampshire

The following answer shows how one student approached the introduction of this report essay. Read it carefully, consider the examiner's comments at the end and think about how the answer could be further improved.

■ ■ ■

Candidate's answer

Introduction

Rural landscapes are complex to define but they are typically classified as less populated parts of the planet with low population densities, and range from wilderness to urban fringe areas.

Leisure and tourism are features of economies globally nowadays and both have been increasing, especially in transition economies such as China.

The question requires an understanding of what is meant by controversy and all that entails, especially conflicts between different users or players. The essay will also discuss how the use of the rural landscape is on the rise and how that too can cause challenges, especially where there is intensive use and the carrying capacity is exceeded. However, management by local, national and international players may reduce the negative impacts.

I will use a variety of case studies to illustrate my answer, including the debate surrounding increased use of Antarctica for ecotourism and the globalisation of golf in, for example, Vietnam. I will also cover smaller-scale examples of conflicts that I have researched, including the use of motocross in Devon (post-productive landscape) and overcrowding of urban country parks. Other controversial issues such as fox-hunting bans in the UK, together with the 2009 Comic Relief Kilimanjaro climb, will be evaluated.

This is a good introduction.

- There is clear reference to the question with the development of a focus.
- There is a section on the framework to be used according to both context (small, large, past and current) and places (e.g. Antarctica, Kilimanjaro.).
- There is less accurate definition of key terms: leisure, tourism, but a good comment on the rise of these activities.

It could have been improved by a link to negative impacts, which create controversies. It should have been written in the third person, not the first person.